DARWIN'S
SANDCASTLE

DARWIN'S SANDCASTLE

EVOLUTION'S FAILURE *in the* LIGHT *of* SCRIPTURE
and the SCIENTIFIC EVIDENCE

GORDON WILSON

ROMAN ROADS PRESS
MOSCOW, IDAHO

Darwin's Sandcastle:
Evolution's Failure in the Light of Scripture and the Scientific Evidence

Copyright © 2023 Gordon Wilson

Published by Roman Roads Press
Moscow, Idaho

RomanRoadsPress.com

General Editor: Daniel Foucachon
Editor: Carissa Hale
Interior Layout: Carissa Hale
Cover and interior illustrations: Joey Nance

Scripture quotations are from The ESV® Bible (The Holy Bible, English Standard Version®), copyright © 2001 by Crossway, a publishing ministry of Good News Publishers. Used by permission. All rights reserved.

All rights reserved. No part of this publication may be reproduced, stored in a retrieval system, or transmitted in any form by any means, electronic, mechanical, photocopy, recording, or otherwise, without prior permission of the publisher, except as provided by the USA copyright law.

For permissions or publishing inquiries, email info@romanroadspress.com

Darwin's Sandcastle: Evolution's Failure in the Light of Scripture and the Scientific Evidence
Gordon Wilson

ISBN 13: 978-1-944482-83-1
Version 1.1.1 October 2023

Creative Commons photo attributions:
Fly Face - Tanel Nook
Russell's Viper - tontantravel
Puff Adder - Danny S.
Malabar Pit Viper - Umakant S Chavan
Copperhead - Edward J. Wozniak

To my older brother and pastor, Douglas Wilson, an unflappable and stalwart man of God who never ceases to fight the good fight. He has been a faithful example of how to gleefully cast down arguments and every high thing that exalts itself against the knowledge of God.

ENDORSEMENTS

"Written in vivid, accessible language, and well-illustrated with drawings and diagrams, *Darwin's Sandcastle* provides a reader-friendly introduction to the major themes and lines of evidence of the Young Earth viewpoint. Dr. Wilson surveys both the Biblical and scientific reasons why someone would defend a young Earth and specially created life. The chapters are brief, but well-supplied with data and arguments. This is a book to give to friends and family who want to understand the Young Earth position. Highly recommended for that purpose."

Paul A. Nelson, Ph.D
Senior Fellow
Discovery Institute

"In *Darwin's Sandcastle*, creation biologist, herpetologist, and filmmaker Dr. Gordon Wilson takes evolution head on using young age creation and intelligent design arguments. There are plenty of creationist books out there with a similar goal. However, one of the problems we have had historically in creationism is that although it was lay ministries and laymen who were brave to take on the challenge of evolution in the last century, few in-depth empirically-based answers were developed to problems evolutionists presented. Wilson's approach has the advantage of several decades of published peer review work from scientists in the intelligent design movement, research Institutes like the Institute for Creation Research and the Logos Research Associates, societies like the Creation Biology and Geological Societies, educational institutions like the Center for Origins Research and Core Academy of Science, conferences like the International Conference on Creationism, and peer reviewed creation journals like the Answers Research Journal. Creationism has been growing up. What Wilson implies in his text in a big way, is that good creation research helps lessen the need for unfounded polemics and points the debate toward the open door of Christian theism."

Joseph Francis, Ph.D
Adjunct Professor of Biology
The Master's University

"Gordon Wilson has put together a content-rich yet approachable book that will appeal to laymen wishing a better understanding of how creation answers our Darwin-saturated world. Light-hearted without sacrificing accuracy, this book defends a solid Biblical worldview without compromise. After having left Darwin's sandcastle washed away by wave after wave of science, Gordon boldly addresses some of the most difficult questions remaining for Biblical creationists. This he does in an even-handed way that teaches the reader how to think rather than what to think. Readers will enjoy Gordon's conversational style, feeling like they are chatting with a friendly science professor at home about matters of mutual interest."

David Coppedge
Science Writer for Evolution News at the Discovery Institute, Former Sys. Admin. Cassini Mission to Saturn, NASA-JPL.

"A cheerful biologist, Dr. Wilson shows his humor and keen views of our world. He presents a creative mosaic from science that shouts 'life is designed.' Insightful, engaging, and joyful, this book points us toward glorious adventures to share with others!'"

D. Eric Aston, Ph.D
Professor of Chemical and Biological Engineering
University of Idaho

CONTENTS

List of Figures	xiii
List of Tables	xiv
Publisher's Preface	xv
Foreword by Douglas Wilson	xix
Acknowledgements	xxv
Introduction	xxvii
Chapter 1: In Whom Do You Trust?	1
Chapter 2: Can Genesis Accommodate Deep Time?	5
Chapter 3: Blind Dating	13
Chapter 4: Designer Genes—What's the Difference Between Microevolution and Macroevolution	29
Chapter 5: Testimony of the Entombed—Does the Fossil Record Reveal Common Ancestry?	55
Chapter 6: Who Was Our Ancestor: *Australopithecus* or Adam?	85
Chapter 7: Alphabet Soup—Can Biological Building Blocks Become a Living Cell by Chance?	99
Chapter 8: Micro-machines—Is a Darwinian Origin of Irreducible Complexities Possible?	115
Chapter 9: Faded Genes—Does Genetic Information Erode?	135

Chapter 10: According to Their Kinds	141
Chapter 11: Biological Badness and the Goodness of God	167
Chapter 12: Tying Off Loose Ends	187
Chapter 13: The End of the Matter	201
Conclusion—A Creationist Manifesto	211
Index	215

LIST OF FIGURES

Figure 1: Coal sample	18
Figure 2: Coal sample after one half-life	19
Figure 3: Coal sample after two half-lives	19
Figure 4: Coal sample after three half-lives	20
Figure 5: A feather	41
Figure 6: A Schizophoran fly face	46
Figure 7: A puparium	46
Figure 8: The Cambrian Explosion	63
Figure 9: The Carboniferous Explosion	66
Figure 10: The beetle fossil record	67
Figure 11: The fossil record of vertebrates	68
Figure 12: The sarcopterygian fish, *Eusthenopteron*	69
Figure 13: *Acanthostega* and *Ichthyostega*	70
Figure 14: *Tiktaalik*	71
Figure 15: The fossil record of amphibians	73
Figure 16: The fossil record of reptiles	74
Figure 17: A simplified family tree of dinosaurs	75
Figure 18: The fossil record of dinosaurs	76
Figure 19: The fossil record of birds	78
Figure 20: The fossil record of various extinct mammal-like reptiles	79
Figure 21: Fossil record of mammalian orders	80
Figure 22: Fossil record of rodent families	81
Figure 23: Fossil record of plant phyla	82
Figure 24: Miller-Urey apparatus	107
Figure 25: Alphabet soup	111
Figure 26: Biceps brachii	121
Figure 27: Section of a muscle cell containing myofibrils	122

Figure 28: A sarcomere   124
Figure 29: Fingers and pens representing a sarcomere   124
Figure 30: Linnaean Lawn   148
Figure 31: Evolutionary Tree   149
Figure 32: The Creationist Orchard   151
Figure 33: Box Turtle Tree   151
Figure 34: Family Emydidae Tree   152
Figure 35: Various vipers and pit vipers   164
Figure 36: Viper venom injection apparatus   173

LIST OF TABLES

Table 1: Selected radioactive elements   17
Table 2: Actual vs. radiometrically measured ages of igneous rockes   24

PUBLISHER'S PREFACE

I remember sitting on the floor of the New Saint Andrews library surrounded by notes from Dr. Wilson's biology class, drilling a classmate on *class*, *genus*, and *species* of various critters. I chose this liberal arts college because I wanted to read great books and grapple with the ideas of the ages past that formed who we are today. What was I doing memorizing that *Diptera* was an order in the class *Insecta*, or that Mitochondria are the powerhouses of the cell? I loved what I was learning, partially because of Dr. Wilson's contagious love for all things living and crawling. But it took me a while to understand why it was part of the curriculum.

The answer lies in the very definition of a liberal arts education. We were not at New Saint Andrews College to become biologists. We were there to understand how God's world works, our place in it, and how to take leadership and dominion in that world. We were there to learn how to be free people.

There is a very real kind of slavery that both the uneducated and narrowly educated share. The uneducated simply don't have the knowledge or intellectual authority to push back against "experts," even when they can feel something doesn't add up. The narrowly educated have what Einstein described as the common condition of seeing a thousand trees, but never the forest, something he observed frequently, even in professional scientists. Because we have abandoned classical liberal arts as a culture, we are largely a nation of uneducated or narrowly educated people. Few of our scientists (Christian or secular) and even fewer of our laypeople have the ability to think in an integrated, cross-discipline fashion. The result is that Christians in particular are unequipped to answer "the specialist." Phrases like "trust the science" or "the science says" hold an illegitimate grasp on the uneducated and narrowly educated alike, and greatly hamper the advance of science itself. A colleague of Gordon Wilson at New Saint Andrews, Dr. Mitch Stokes, explores the pitfall of narrow, specialized education in an essay for Roman Roads Press titled "What does Jesus have to do with STEM?" which I recommend to anyone seeking to further understand this concept.

    The long term solution to this problem is the renewal of classical learning at a generational level grounded in an evangelical and lively faith. But in the short term, books like the present provide a kind of layman's handbook to the big picture on this topic which is closely guarded by "specialists." See the forest and not just individual trees, understand the big picture of the arguments and science involved in the Creationist position, confident in your faith and convictions. Gordon Wilson has dedicated his life to helping people navigate in a world of scientific idolatry, not by being anti-science but on the contrary by teaching good science at an approachable level.

As someone who personally knows the author, studied under him, experienced his immense love for God's creation, watched him debate Evolutionists, and encourage and admonish Christians who disagree on these subjects, I can think of no better person to write this book. If you want to hear his voice and see the joy he takes in God's Creation, I also encourage you to watch the *Riot and the Dance* nature documentaries which he narrates.

Daniel Foucachon, Publisher
September 2023

FOREWORD

Conservative Christians frequently go through three stages in their attitude toward the theory of evolution. The first is when they are young students, encountering the idea for the first time, and in that initial encounter they think it is the dumbest thing they ever heard. They have this reaction because they are being brought up in a Christian "plausibility structure" that rejects evolution, and because they can also see the similarity between evolution and Kipling's "just so" stories. They see the problems with evolution, but it is also true that they live in an environment that makes it very easy to see those problems—and where you might get hooted at by your peers if you didn't see the problems.

The second stage happens when they encounter their first intelligent evolutionist, perhaps in a book, or with a professor at college. The safety of the plausibility structure is now gone, and it is evident to the young person that not only is this "engaging and funny biology instructor" not a perfect moron, he is also the master of more scientific knowledge than this young naïf—a history major—will ever hope to possess. The result of this, at a minimum, is that our student is rattled. He may even start looking around for ways to "harmonize" what he thinks he is learning with the faith he grew up with, and when that starts to happen, it is the faith that has to do all the stretching.

The dilemma at this stage is caused by a phenomenon that Gordon Wilson acknowledges at the front end of his argument. As Wilson notes, many advocates of evolution are learned, careful, intelligent, and competent in their respective fields. At the same time, the central pillar of their scientific worldview—materialistic evolution—is simply unreasonable. It beggars belief, and yet many educated people believe it. Not only do they believe it, but they vigorously police the borders of *their* plausibility structure—informing us that no respectable scientist disputes evolution, while carefully defining respectability as requiring a belief in evolution.

It really does present a discordant picture, and it is little wonder that our student was thrown by it initially. This is a bizarre inversion of the idiot savant—someone who is seriously disabled in most areas, but who operates at genius levels in one area. Here is the inversion. In almost every area of life they are bright, educated, accomplished, and yet in this one area of their worldview, they stoutly maintain that triangles have five sides. How to account for this?

Our student is not going to break through to the third level of awareness unless and until he comes to see that the biblical worldview not only accounts for the exquisite engineering of the falcon's eye, but also for the stubborn blindness of the scientist studying the falcon's eye. The Christian faith accounts for what the falcon can see and what the scientist cannot see. Both phenomena require an accounting; both demand an explanation. We should stagger under the weight of two things—one being the wisdom and knowledge of God, and the other being the mystery of lawlessness.

Scripture teaches us that it is possible to be very clever and also to be a rebel against reality. The biblical worldview teaches us that man was created to exercise dominion over the world, which includes investigating and understanding it, and the Word also teaches us that we rebelled against the only true foundation of our appointed station. This means that to acknowledge creation means to acknowledge a Creator. And acknowledging a Creator amounts to acknowledging the existence of an unbeatable rival. And we apparently can't have that. We are too clever to tolerate anything of the kind.

And so, to preserve the unquestioned dignity of our manifest cleverness, we have established a manifest absurdity as the central dogma of the modern age. The leap across the divide between inorganic and organic is the miracle of miracles, made all the more marvelous by the fact that this miracle happened all by itself, and with no miracle worker to make it go. We then think to make this initial unbelievable thing more palatable to our insulted common sense by multiplying an inverted pyramid of unbelievable things that somehow cascade up the endless expanding staircase of blind happenstance. May as well believe that your kid tripped while putting away his Lego set, only to watch in astonishment as it assembled itself into a working model of a nuclear reactor. And don't tell me that we need deep time for such accomplishments because even with deep time, on average the evolutionist needs a grand miracle every thirty seconds or so.

Look. I think my illustrations are outlandish also, but what is more complicated than a Lego reactor? I don't know—a hummingbird's heart, a butterfly's antennae, fifteen things that your liver is doing right now, a wombat's nose, an armadillo collapsing into a ball, and the engineering of your toddler's wrist. If we had unlimited funding, we could spend the rest of our lives finding more complicated systems than a reactor within one square foot of my front lawn.

In this book, Gordon Wilson carefully walks the reader through the basic arguments against evolution. But he does this within a framework of the entire Christian worldview. Whether mindless evolution occurred or not is not a detail. It is not a dispute about whether a particular incident happened within this world. If Smith ordered a mocha this morning at a coffee bar is an issue that, whether it happened or not, would leave the world pretty much the way it was before. But if evolution occurred, the world is one kind of place, and if it did not, it is completely another. So the issue is necessarily an ultimate one—and concerns what kind of world this even is.

Gordon understands particular arguments against evolution must be presented within this larger framework. But with that understanding, he works through the problems with radiometric dating, the gulf between inorganic matter and life, the real meaning of the fossil record, and the staggering details involved with irreducible complexity, and more. The arguments are preeminently reasonable, but it still takes faith to follow them. This faith is not to fill up any deficiencies in the arguments, but rather to address the deficiencies created in our reasoning by our rebellion against God. Read, and learn. Read, and be fed.

One last comment. It is a privilege to write a foreword to a book written by my brother. I don't get enough opportunities to say how proud I am of his scientific accomplishments, coupled together with his robust biblical faithfulness.

Douglas Wilson
September 2023

ACKNOWLEDGEMENTS

I would like to express my gratitude to the many academic folks in the young Earth creationist community who have blazed the trail in the popular and technical arena that are faithful to Scripture and science (rightly understood). I am also grateful to those in the Intelligent Design movement who have presented excellent arguments that clearly expose the bankruptcy of Darwinism.

Many thanks to Daniel Foucachon of Roman Roads Press, who was eager to publish this book. I am very grateful to Carissa Hale (Roman Roads), who painstakingly edited the book, and my friend Eric Aston, who volunteered to edit the book with an eye on grammar and scientific accuracy. A big thanks goes to Josiah Nance for his cover design and the many illustrations needed in the book's innards.

Lastly, I thank my wife, Meredith, for her unwavering support and loving encouragement to keep my hand to the plow in teaching and writing.

INTRODUCTION

Countless books have been written on the subject of evolution and creation over many decades, by Darwinists, disillusioned Darwinists, creationists, and Intelligent Design (ID) advocates. While some have been written in a polemical style, the current trend is towards a sterile scientific style preferred by science nerds (full of insipidly accurate, thoroughly referenced, technical, and objective arguments against evolution) in order to get a hearing from respectable, academic types. This isn't necessarily a bad thing, but authors often think they can persuade by the force of reason and evidence alone. After all, shouldn't reasonable evolutionists cave to the weight of evidence and logical argumentation and cry, "What must I do to be saved?" For many years I thought this approach should have that kind of effect but alas, it did not. Why? There is one major reason it doesn't. Most evolutionists are not reasonable on the ultimate question, "What is the ultimate cause of the reality, complexity, unity, and diversity of the universe?" On that question, unbelieving scientists, no matter how intelligent, are simply not reasonable, they are foolish. I don't call them fools on my own authority;

the Bible does (Psalm 14:1; Romans 1:22). Anyone who thinks that hydrogen gas (from a point singularity which sprung from nothing and exploded in the Big Bang), given 13.8 billion years, can evolve into the present universe containing planet Earth, on which millions of complex species (including humans) developed, is foolish. And when these humans (who design and build iPhones, rockets, and satellites, who transplant organs, walk on the moon, write symphonies, paint and sculpt masterpieces, etc.) speculate on how mindless, purposeless matter evolves into all this stuff along with all our accomplishments, they are simply not being reasonable.

But the reason why creationists so often try to appeal to an evolutionist's reason is that, in one sense, evolutionary scientists are usually both exceedingly reasonable and intelligent when practicing the scientific method on empirically answerable questions. Just look around and see all the great accomplishments of the modern scientific community: medical breakthroughs, space travel, nuclear power, computer technology, DNA technology, etc. These successes weren't the work of idiots. Many secular scientists are masters of their craft and demonstrate their intelligence through brilliant experimental design and logical problem-solving skills when they get the right answers over and over again.

Creationists have for decades been scratching their puzzled noggins, asking themselves, "Why are these brilliant and logical scientists not reasonable on this ultimate question?" We wrongly think that biblical creation can be deduced by empiricism and logical syllogisms without revelation, faith, and repentance granted from God. If it can be deduced this way, then evangelism should start with teaching introductory logic to the target audience. A corollary to this is that only smart, logical people will be saved by coming to the Father through their intellect by rigorous study (not through faith in Christ) and that slow, stupid, illogical people will go to Hell, because they don't have the brains or wit to reason out the truth of God and His gospel. No, evangelism should start with the Gospel preached to everyone, the wise and the foolish, Greeks and non-Greeks (Romans 1:14). These truths are logical, but faith is not coming to grips with overwhelming scientific evidence. Faith isn't working out the syllogism correctly. These truths are spiritually discerned by faith. The atheistic scientist, no matter how intelligent, is spiritually blind and must first receive spiritual sight from the Holy Spirit. If this does not happen, any persuasion will not lead to salvation unless God grants faith. It may result in intellectual assent that there may be an intelligent designer of some sort (a la Anthony Flew, Thomas Nagel, or David Gelernter), but they are not half saved if they get to that point, and we shouldn't pat ourselves on the back because our apologetics made an atheist into an agnostic or an unregenerate theist or deist.

If I can't reason them into faith in God and the gospel, why am I writing? My goal in writing this book is twofold. First, I find that believing Christians (whose faith in the Bible is sorely tried by the incessant secular narrative thrown at them from every quarter) are greatly encouraged when they see that the evidence, rightly understood, bolsters their faith in the same way Thomas's faith was strengthened when he placed his fingers in the nail prints of Jesus' hands.

Secondly, it is to demonstrate to unbelievers, who have placed undue faith in secular science, that naturalistic evolution is only impressive to their uninformed imaginations. In reality, it's a sandcastle (made to look like an imposing fortress), and the builders are hoping their moat will stop the incoming tide. I want to show them in no uncertain terms that to cling to their pitiful theory is to embrace an irrational faith.

My goal is to encourage sincere Christians who may have begun to doubt the historicity of Genesis and have begun to lose their footing as they wade against the strong current of unbelief masquerading as science. When John the Baptist sent his disciples to check to see if Jesus was the Christ, Jesus told him what he needed to hear. He didn't get chewed out for doubting. In the same way, I want to recount biblical and scientific truths using accessible language to reveal 1) the historical reading of Genesis, 2) God's glory as shown in the creation, and 3) Darwinism's bankruptcy (2 Corinthians 10:5) with my own reflections and insights worked in. I also want to persuade, but that is only desirable if one's thesis is true and convinces more people of the truth. If it isn't true, then persuasion simply draws more people under one's false teaching, which will be judged more strictly (James 3:1). Both truth and persuasive writing should be used in concert in "proving that Jesus is the Christ" and that evolution is a deceitful and destructive false doctrine that diverts human praise toward mutation, natural selection, etc., rather than to the Almighty God. Because the prophets, the apostles, and Jesus Christ all effectively used both truth and persuasive speech (and writing) in preaching the Kingdom of God (which presupposes the truth of God's Word), I will strive to do so as well, knowing that both are fruitless unless the Holy Spirit makes the heart receptive.

I want to demonstrate to my readers the utter failure of Darwinism. I want to do my part in demolishing this insidious stronghold (2 Corinthians 10:4–5) so that no one will be taken captive by this hollow and deceptive philosophy (Col. 2:8). I also hope to redirect glory and thanks for the creation toward Jehovah, Who alone is worthy. In our Darwin-dominated society, blind chance, mutation, and natural selection have received most of the glory for the unity, diversity, and complexity of the natural world. It's about time this philosophy is seen for what it is: an emperor that has no clothes. Maybe the public will be emboldened to rally round and point the finger of scorn.

CHAPTER 1
In Whom Do You Trust

There are many flavors of theologically confused or compromised evangelical Christians regarding Genesis 1–11 (theistic evolutionists, progressive creationists, day-agers, gap theorists, frameworkers, etc.). These folks usually claim to hold a high view of the authority of Scripture but when we look at how they interpret Genesis 1–11, it is pretty obvious who or what they ultimately trust. I have heard some say that they hold to one or more of the above positions for exegetical reasons and could care less what the secular scientific community says. For the record, I am exceedingly dubious of that claim. They may not be deliberately lying. They may be sincere, but I don't think they realize how much the secular narrative of world history has permeated their very bones. They know (deep down) that holding to a young-Earth position is tantamount to alienating yourself from any hope of academic respectability. They know that secular academics don't usually think of young-Earth creationists as bright people sadly mistaken; rather they usually think young-Earth creationists are a less-evolved life form. How can you cultivate an intellectual persona while holding to a position that seems to be on par with the Flat Earth Society or the Alien Abduction Survivors Network? If Scripture seems in conflict with what "science" says, then in their mind "science" is the rock and Scripture is the putty that must be molded to fit the "scientific facts." If they feel the need to consider Scripture authoritative in some way, they find a "respectable" theological camp that reinterprets Scripture so that it comfortably conforms to the prevailing views of "science."

However, Scripture isn't Silly Putty® that is to be sculpted to fit the current views of secular science (as BioLogos would have us believe). Scripture must be interpreted carefully using responsible hermeneutic principles by scholars who couldn't care less if their interpretation doesn't jive with current secular scientific proclamations. Why are some evangelical pastors and theologians fearful of believing in a literal Adam, a global flood, and a relatively short time frame for all cosmic history? Do they really think it is untenable exegetically? They shouldn't. Here is a very revealing quote from the world-renowned Oxford Hebrew scholar, the late Professor James Barr:

> ...so far as I know, there is no professor of Hebrew or Old Testament at any world-class university who does not believe that the writer(s) of *Genesis* 1–11 intended to convey to their readers the ideas that: (a) creation took place in a series of six days which were the same as the days of 24 hours we now experience; (b) the figures contained in the *Genesis* genealogies provide by simple addition a chronology from the beginning of the world up to the later stages in the Biblical story, and (c) Noah's flood was understood to be

worldwide, and to have extinguished all human and animal life except for those in the ark.[1]

It is sadly the case that James Barr and the scholars he's referring to don't believe Genesis 1–11 as containing true historical events. Why? Because they trust in secular science instead. But one thing these professors are certain of is this: the writers of Genesis 1–11 thought they were recording true historical events and were not trying to communicate something compatible with the secular scientific narrative, i.e. the Big Bang, origin of life, biological evolution, etc. They know what the writers of Genesis intended to mean. They know secular science's view of origins isn't even close to being in the same book, let alone on the same page, as Scripture. In whom do we trust? The Word of God, Who was there, or the word of man, who was not there? Scientific claims must be interpreted in light of Scripture; not the other way around. In the next chapter, I hope to summarize why Genesis 1–11 is a historical narrative and should be believed in the same way we believe the events contained in the Gospels or any other historical part of the Bible.

---

1 Cited by Alvin Plantinga in "Evolution, Neutrality, and Antecedent Probability," in *Intelligent Design Creationism and Its Critics*, ed. Robert T. Pennock (Cambridge, MA: MIT Press, 2001), 217. From a letter to David C. K. Watson (April 1984). See also Don Nield, Re: [asa] The Barr quote, May 16 2008, accessed April 4 2023, http://www2.asa3.org/archive/asa/200805/0399.html

CHAPTER 2

Can Genesis Accommodate Deep Time?

There are two main biblical arguments that preclude the shoehorning of deep time (millions to billions of years) into the first eleven chapters of Genesis: 1) the meaning of *yom* and 2) the genealogies of Genesis 5 and 11. But before I discuss them, I want to say a few words about our spiritual state before we enter into a discussion about the age of the earth. Any debate on this touchy matter should be done in a gracious manner. We should never bluster. "As far as it depends on you, live at peace with everyone" (even in disagreement) (Romans 12:18). Also, "everyone should be quick to listen, slow to speak and slow to become angry (James 1:19)." A Christian should have a gentle (not wimpy) demeanor. This is achieved by being filled with love, joy, peace, patience, etc. Getting a bumper crop of the fruit of the Spirit is obtained by confessing your sins to God with a humble and contrite spirit. This is not an academic exercise at all and consequently those who fancy themselves intellectuals often overlook it. Persuasion is not just a function of the soundness of the arguments;

it is also a function of the spiritual soundness of the one presenting them.

Let's move on to the arguments.

1) What does *yom* mean? *Yom* is the Hebrew word for "day" and has a similar semantic range as the English word *day*. It doesn't always mean a 24-hour solar day. It can mean just the daylight portion of the 24-hour period, or it can mean a generation or so, as when your Grandpa says, "in my *day* (generation) we put in an honest *day's* (daylight period) work." However, in the Old Testament the overwhelming majority of the time it means a regular solar day. In the days of the creation week, numerical modifiers like second, third, etc., along with the phrase "evening and morning" are ways that a Hebrew author would narrow the meaning to a literal solar day. There are two passages containing the word *yom* where one of these modifiers is used and is not necessarily to be taken as a literal day. These passages are Hosea 6:2 and Zechariah 14:7. However, they have a prophetic context which clues in the reader to take it figuratively. The context in Genesis, however, is historical narrative. How do we know Genesis is a historical narrative? One way is to look at how the New Testament writers referenced it. Jesus in Matthew 19:4 says, "He who created them from the beginning made them male and female." This indicates that Adam and Eve were real historical people and not a mythical couple useful for teaching about the institution of marriage. Paul says in Acts 17:26a, "And he made from one man every nation of mankind to live on all the face of the earth…" Here Adam is referred to as the ancestor of all mankind and he is certainly not a mythical ancestor. Does historical narrative mean there can't be literary structure?

Absolutely not. Even though most of Genesis is straightforward historical narrative, there are a variety of literary structural elements embedded in it; but it is at least to be taken as history in the real past. Old Testament scholars and theologians may find all sorts of interesting typologies that foreshadow future events or people. There are chiasms, poetic elements, parallelisms, and other deeper layers that underscore the richness of God's Word. Nevertheless, these things were not intended to negate the plain historical meaning that a child or a lay person can grasp.

What was its intended meaning? Most people reading Genesis 1–11 don't walk away thinking that it spanned deep time. Translators chose the English word *day* for a reason. If the Hebrew authors wanted to convey the idea of vast amounts of time compatible with the secular geological narrative of Earth history, they have other, more appropriate words at their disposal. For instance, *olam* means a long duration of time. Why not use that word if you want your readers to grasp the immense age of the earth? A plural form of *yom* like *yamim rabbim* means "many days, long life" but not deep time.[2] If secular scientists are correct about the age of the earth, and if the writers of Genesis are wanting to record an accurate historical account, then they did not choose the right word or words to communicate the idea of deep time.

Some may object, saying that Genesis is not a science textbook. Young-Earth creationists don't claim it is! History books can communicate the idea of deep time. I don't know Hebrew, but I do know that King Asa went

---

2   BibliaTodo Dictionary, Bibliatodo, accessed on April 4, 2023, https://www.bibliatodo.com/en/bible-dictionary/yamim-rabbim.

against an Ethiopian army of a million in 2 Chronicles 14:9, so apparently big numbers can be communicated in the Hebrew language. There is also *shanah*, a Hebrew word for year. *Shanah* coupled to large numbers used to quantify the aforementioned Ethiopian army could really get at the idea of deep time. If kids can grasp the idea of deep time in popular dinosaur books, certainly ancient Israelites were astute enough to do the same when reading Genesis. But deep time was not written or intended by the authors of Genesis and we need to come to grips with that fact. Why are we tempted to believe convoluted and nuanced old-Earth interpretations that almost require a graduate degree in Theology just to get our head around them? Why not believe the most straightforward interpretation? I think it's because the propaganda machine of secular science has successfully inculcated the notion of deep time into the noggins of many evangelical theologians and Bible scholars.

2) I will address the genealogies of Genesis 5 and 11 next. This will preclude any other attempts to shoehorn deep time into the first eleven chapters of Genesis. Once we have a pretty good notion of its time frame, we can then move on to the bluster of the secular scientific community that has intimidated biblical scholars and lay Christians for far too long. Many feel they can't commit to a young-Earth position because they think it's an academically indefensible position in light of the "overwhelming evidence." Does the secular world really have young-Earth creationists in checkmate, or do we just believe them when they say, "checkmate"?

I mentioned above that *yom* in the context of Genesis 1 was a regular solar day. To say the authors intended to convey eons of time is simply irresponsible exegesis. In short, the creation week was just…a week. So where else can theologians and clergy (who are under the spell of secular scientific consensus) add that precious time to Genesis? Even though many old-Earth creationists realize that their desired deep time is best sought for before day 6, many have still tried to insert more time in the genealogies of Genesis 5 and 11. Even if honest exegesis allowed it, it won't come close to adding the needed geological time demanded by evolutionists. At best, it would bring them into conformity with some of the dubious secular chronologies of other ancient civilizations. But does the text allow us to shoehorn several thousand or more additional years into the genealogies between Adam and Abraham? No.

To prove it to yourself, get some graph paper. Lay the paper in the landscape position and at the bottom of the sheet label the X-axis as time where each square represents 100 years. Make a column on the left margin of the sheet and list the men in the genealogies starting with Adam at the top, Seth right below Adam, Enosh right below Seth, Kenan below Enosh, and so on. Just to the right of this column of names draw a vertical line (like a Y-axis). This represents Time Zero when Adam was created. Now here comes the fun part. With a colored marker draw Adam's lifespan (moving right from Time Zero) as a horizontal bar. Since Adam lived 930 years, his bar should cover 9.3 grid squares (since each square is 100 years). In Genesis 5:3 it says, "When Adam lived 130 years, he fathered a son in his own likeness, after his image, and named him Seth." So now on the Seth "row" right below Adam, draw (with the marker) Seth's lifespan bar starting at when Adam was 130 years old. On the graph, Seth's bar starts right at the 1.3 mark of Adam's life and continues for 912 years. Now do Enosh. He was born after Seth lived 105 years, so draw Enosh's lifespan bar starting a smidge more than one square (105 years) after Seth was born. If you follow my directions correctly and continue to make overlapping life span bars all the way to Abraham, you'll find that the genealogy is seamless. In other words, there isn't a way to squeeze X amount of time in between these men.

One might say that Seth wasn't a direct son of Adam or that Mahalalel wasn't a direct son of Kenan. Maybe they were grandsons, you say, or great-grandsons or great-great-grandsons. Do I hear three 'greats'? For the sake of argument, let's say there were several missing generations between all the men on the list. Maybe there were men that, for whatever reason, the author of Genesis decided to leave out. Fine. That doesn't change the timeline at all. Why? Quite simple; the author gives the ancestor's age when he begat his next recorded descendant. So even if there were several missing generations between say Kenan and Mahalalel, Kenan was still 70 years old when he begat Mahalalel. Whether Mahalalel was his son, grandson, or great-grandson, it doesn't make a lick of difference to the timeline. The only discrepancy is that Genesis 11:12–13 has no Cainan (Kenan) between Shelah and Arpachshad whereas the Luke account does. This could be a scribal copying error or there was another fellow named Cainan that Luke put in and the author of Genesis leaves out. Either way, Arpachshad was 35 when he begot Shelah, so again the timeline remains un-stretched. The bottom line to all this is that Abraham was born about 2,000 years after creation. Attempts to make the timeline stretch to fit the various secular chronologies is simply mishandling Scripture. If we put a date on Abraham, then we necessarily put a date on Adam and consequently creation. Although there are minor discrepancies on when Abraham was born, most Old Testament scholars place him around 2,000+ BC. So, you do the math.

Some might ask what the consensus of the church fathers was regarding the age of the earth. Overwhelmingly the church fathers believed in a young Earth. Thankfully, they didn't have the modern scientific community breathing down their necks. Some, ironically, use St. Augustine's view as precedent for not taking the creation week literally so that they can feel justified in taking an old-Earth view. Yes, St. Augustine wasn't literal but he actually believed that God created *instantly* and revealed it over the space of a week for our understanding, not over vast geological ages.[3] So, to use St. Augustine to promote an old-Earth view is disingenuous because he actually believed in a slightly young*er*-Earth than typical young-Earth creationists. If you want to take a deeper dive into what St. Augustine and other church fathers believed about Genesis 1–11, the church's compromise with deep time, the global flood, etc. from a historical and exegetical approach, *Coming to Grips with Genesis* is a must read.[4]

---

[3] St. Augustine, *The Literal Meaning of Genesis*, trans. John H. Taylor (Westminster, MD: Newman Press, 1982).

[4] Terry Mortenson and Thane H. Ury, eds., *Coming to Grips with Genesis* (Green Forest, AR: New Leaf Publishing Group, 2008).

CHAPTER 3
Blind Dating

Despite the valiant efforts of theologians and Bible scholars to shoehorn deep time into the Bible (because they want to "follow the science"), it would be good first to see if "the science" (e.g., radiometric dating methods) is trustworthy and produces indisputable facts about the age of the earth. If they aren't trustworthy, the efforts of past and present theologians to accommodate deep time were and are unnecessary.

My dad told me that when he was a midshipman at the Naval Academy (class of '50), some midshipmen practiced an ungodly and cruel tradition. Unattractive girls were referred to as "bricks." If a midshipman caught another midshipman dating (what he deemed to be) a "brick," he would give him a left-handed salute. This underhanded insult (the girl was oblivious) would push all the wrong buttons on the midshipman, but I digress. I don't want to discuss the topic of dating bricks. However, in discussing evolution, the topic of dating rocks almost always comes up. So why did I entitle this chapter, "Blind Dating"? In the colloquial sense, blind dating has more in common with radiometric dating than you might think. Both kinds of blind dating are subject to high levels of cluelessness. The consensus view that the universe, Earth, and life are billions of years old is often used as a trump card against creationists. The idea of deep time has been so thoroughly ingrained into our collective noggins, that to say otherwise can get you instantly pigeon-holed as a hick, bumpkin, cornpone, a member of the Flat-Earth Society, or all of the above. In other words, it is not an easy task being taken seriously after you identify yourself as a young-Earth creationist. Nevertheless, you should embrace the title, not shy away from it. But when you do, it behooves you to know what you're talking about both scripturally and scientifically. Once you have, you will find a wonderful confidence welling up inside. Knowing what you're talking about makes you less likely to act like a cat cornered by a pack of dogs. Many Christians have acclimated to living within a Christian bubble and don't bother equipping themselves with answers about the age of the earth. If they encounter someone who differs and who knows their stuff, they often crumple under the other's arguments. They were caught off-guard without a glimmer of a counter argument. This is a tragedy. I hope to not only strengthen your faith in the historicity of Genesis but also turn the tables so that your opponent is the one who feels embarrassed. We must also keep in mind that it is really a separate topic from the theory of evolution. Deep time (hundreds of millions of years) is of course a necessary condition for neo-Darwinian evolution, but belief in deep time doesn't necessarily imply evolutionary thinking. There are creationists who believe in deep time and yet reject evolution. Therefore, it needs to be addressed as a separate matter.

As argued in Chapter 2, the notion of deep time cannot be reconciled with any responsible exegesis of Genesis 1–11. If Genesis precludes deep time, then how do biblical creationists deal with the supposed overwhelming scientific evidence that the universe, Earth, and life are billions of years old? I'm glad you asked. To answer that question, we need to examine our starting assumptions about how scientists date things. Let's start with rocks (not bricks).

Geologists who specialize in dating rocks and minerals are called geochronologists. In order to evaluate the certainty of the dates they assign to rocks, one doesn't have to know everything about the procedure. What one needs to know are any assumptions they make about the rock sample they are dating. This has been discussed countless times in the creationist literature, but I hope to make it clear and simple without getting bogged down with unnecessary details.

RADIOMETRIC DATING

Radiometric dating is a technique that considers the percentages of radioactive *parent* and its non-radioactive *daughter* elements in a rock sample. First, you might want to know what *parent* and *daughter* means in this context. *Parent* refers to the radioactive element that is in the rock. For example, uranium-238 (U^{238}) is a radioactive element that is found within certain rocks. The atomic number of any element is the number of protons it has per atomic nucleus. It is an element's identity. Hydrogen's atomic number is 1, helium's is 2, carbon's is 6, oxygen's is 8. Uranium, a much heavier element, has an atomic number of 92. In U^{238}, the 238 refers to the atomic mass (the sum of protons and neutrons per nucleus). So in the atomic mass of 238, 92 are protons and 146 are neutrons.[5] U^{238} undergoes radioactive decay by spitting out alpha particles every so often. Each time it does so, it not only loses mass, it changes its elemental identity. An alpha particle is two protons and two neutrons. One alpha particle lost means that the atom is no longer uranium; it is now thorium (atomic number 90; also radioactive). Let's cut to the chase. In the radioactive decay of uranium, it slowly loses several alpha particles until it becomes the stable, non-radioactive element of lead, Pb^{206}, as the final daughter element.

Half-Life

Various radioactive elements decay at different rates. The term *half-life* refers to the time it takes for one half of the parent atoms to decay into the daughter atoms. You might think that after the second half-life, the remaining parent would become daughter, but that is not the case. Instead, during the second half-life, *half of the remaining parent* will decay into daughter. In other words, after two half-lives ¼ will still remain parent and ¾ will now be daughter. After three half-lives, ⅛ will still be parent and ⅞ will be daughter

---

[5] The number of neutrons can vary. An element with a variable number of neutrons is called an isotope. Example: U^{235} is an isotope of uranium with 143 neutrons. If it is uranium, it will have 92 protons.

and so on (half of ¼ is ⅛). Below is a table showing the half-life of several radioactive elements.

Parent	Daughter	Half-Life
Uranium^{238}	Lead^{206}	4.5 billion years
Carbon^{14}	Nitrogen^{14}	5,730 years
Potassium^{40}	Argon^{40}	1.28 billion years
Rubidium^{87}	Strontium^{87}	49 billion years

Table 1: Selected radioactive elements. Each row shows the radioactive parent, the daughter, and the half-life.

Let's use carbon-14 (C^{14}) as a simple example, since it has a relatively short half-life and is often useful for dating objects that once lived. Assume we have a coal sample, created from dead plants and animals. The vast majority (99%) of all the carbon on Earth, including this sample of coal, will consist of the non-radioactive carbon C^{12}. Since C^{12} is the most abundant stable isotope, it serves as a practical internal reference for age-dating. C^{13} also exists but it is non-radioactive and only about 1% of Earth's carbon, thus it is less useful and often ignored. C^{14} however is radioactive and found in trace amounts. What geochronologists look for is this rare radioactive C^{14} which decays into nitrogen-14 (N^{14}). Now suppose when the coal was formed, the parent-daughter percentages were 100% parent C^{14} and 0% daughter N^{14}. See Figure 1 below. White represents C^{14}.

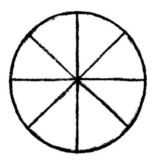

Figure 1: Coal sample—100% parent C^{14}.

Let me clear up some possible confusion, since you may have noticed that the atomic mass of 14 didn't change from C^{14} to N^{14}. In contrast, U^{238} undergoes alpha decay where significant mass is lost each time an alpha particle is ejected (minus 4!) until it finally becomes the much less massive Pb^{206}. Not so with C^{14}. It has six protons and eight neutrons, hence an atomic mass of 14. C^{12} has six protons and only six neutrons, and C^{13} has six protons and seven neutrons. C^{14} undergoes *beta* decay. In other words, one of the neutrons spits out an electron (or beta particle) to become a proton. Consequently, the atom is no longer carbon-14 but it is essentially the same atomic mass as a different element, a stable form of nitrogen (N^{14}) now with seven protons and seven neutrons.

What would the parent-daughter percentages be after 5,730 years (one half-life)? Answer: 50% C^{14} and 50% N^{14}.

The figure below represents these percentages after one half-life. The white again represents the parent (C^{14}) and the shaded represents the daughter (N^{14}).

Figure 2: Coal sample after one half-life—
50% parent C^{14}, 50% daughter N^{14}.

After **two half-lives**, it would be represented like this:

Figure 3: Coal sample after two half-lives—
25% parent C^{14}, 75% daughter N^{14}.

And after **three half-lives**, like this:

Figure 4: Coal sample after three half-lives—
12.5% parent C^{14}, 87.5% daughter N^{14}.

So after the coal sample has been sitting around for three half-lives (5,730 x 3 = 17,190 years), ⅛ is still parent and ⅞ has decayed into daughter and therefore it is given a date of 17,190 years old. As you can see from these illustrations, a geochronologist assumes a few things. He assumes the relative percentage of the parent (and sometimes of the daughter) when the rock was formed by comparing it to a similar reference sample with a known date of death (or date of harvesting for plant materials). In this case, the most direct measurement is a ratio of the C^{14} to the C^{12} (as the internal reference for percentages and ratios), since the half-life calculation does not actually require the amount of N^{14}. He also assumes that nothing other than natural radioactive decay has altered the percentages of these elements during its unobserved history and that the rate of radioactive decay never changes. The same basic measurements of relative changes in parent-daughter ratios (relative to each other or another common reference, as also with isochron dating discussed below) are determined in all the many radiometric dating methods. The differences lie in the radically different half-lives and the identities of the parent-daughter pairs.

Let's consider the legitimacy of these assumptions that the geochronologist made about the rock. It should be obvious that he wasn't around when the rock was first formed, so he doesn't actually know the initial ratio of parent to daughter(s). What if some "daughter" was there at the start?[6] That would yield an older inferred age from time zero, when the rock was made. He doesn't know what occurred to the rock during all the time it was sitting around (nor what occurred to the material going into the rock before the rock was formed). Cosmic rays, ground water, etc., could have passed through the rock during its unobserved history, adding or subtracting from the parent or daughter and thus messing up the ratio. Lastly, he doesn't know if the radioactive decay rate has been constant during the rock's history. Assuming a constant decay rate may seem more reasonable than the first two, but there is good reason to doubt even this assumption as well. Let's list these assumptions more concisely.

1. The parent-to-daughter ratio (or percentages of each) is known at time zero.

2. The rock has remained in a closed system during its unobserved past. Nothing was added or subtracted that may have altered the ratio. Only radio-

---

6 In Carbon-14 dating, Nitrogen-14 is often ignored.

active decay is assumed to alter the amounts of parent and daughter.

3. The decay rate has remained constant during its unobserved past.

There are very strong reasons not to trust these assumptions.

1. There is really no way to know what the ratio of parent to daughter (or percentages of each) was at time zero. That assumption could easily be wrong for a number of reasons. *To use an hourglass analogy, to know how long the sand has been trickling, you have to know how much sand was in both the top and bottom of the hourglass at time zero. The top sand represents the parent and the bottom sand represents the daughter.*

2. Rocks are not in closed systems. Various things (particularly water) can leak through rocks which can subtract parent or daughter or add false parent or false daughter (i.e., the same elements that weren't part of the decay process). *Unlike the hourglass, rocks aren't closed systems. You can't tell how long it has been trickling if sand was added or subtracted to the top or bottom during the unobserved past.*

3. There is good evidence that decay rates may have been different in the unobserved past. *In the hourglass analogy, the trickling rate may have been faster (or slower) in the unobserved past.*

A sophisticated method called *isochron dating* has been developed to make the first assumption unnecessary because it supposedly can determine the daughter abundance at time zero. It also claims to identify any changes due to open-system disturbances. However, isochron dating isn't all that it's cracked up to be. Geochronologists have checked its accuracy by using it on igneous rocks of known ages, and it has been shown to be in great error. For example, two east African lava flows of historically recent volcanic activity were dated using the isochron method and yielded a date of 773 million years old. When the actual age of the lava rock is known and the isochron (or some other) method yields a date that's extravagantly wrong, something is terribly amiss.

Geochronologists are quite adept at figuring out all the factors that contributed to the massive error but this doesn't absolve them of all wrong. If dates are often frightfully wrong for rocks of known age, why should they be trusted on dates assigned to rocks of unknown history and age? Tragically, their dates are accepted as "gospel." The table on the following page contrasts the actual age of the rock unit (volcanic rocks of recent origin) to the "age" based on some form of radiometric dating.

Some claim that rocks known to be very recent, i.e. younger than a certain age, should not be dated using these methods as they will yield inaccurate results. If that is true, why are rocks of unknown age assumed to be old enough to justify the particular dating method? Maybe they are just as young as the rocks they say shouldn't be dated.

Location	Actual age based on historical records	Measured age based on radiometric dating
Hualalai (Hawaii) volcanic basalt	1800–1801	22.8 million years old
Kilauea (Hawaii) volcanic basalt	Less than 200 years ago	21 million years old
Mt. Stromboli, Italy	1963	2.4 million years old
Medicine Lake (Montana) obsidian	Less than 500 years ago	12.6 million years old

Table 2: Actual ages of several igneous rocks compared to their radiometrically measured ages (data from mainstream journals).

Problems with C^{14} Dating

Rocks thought to be millions of years old are considered too old to be dated using C^{14}. Because the C^{14} half-life is relatively short (5,730 years), it is considered inappropriate for any rock assumed to be older than 60,000 years. The reason for this is that after ten half-lives, there are so few remaining C^{14} atoms that they are too difficult to detect with accuracy. C^{14} dating is generally used for human archaeological artifacts and other carbon-containing material that is assumed to have originated from the recent past (several thousand years ago). However, many creationists don't think anything is much older than 6,000 years. With this biblical assumption, a team of ICR (Institute for Creation Research) scientists studying radioisotopes and the age of the earth (or RATE) decided to date some coal samples. Although the absolute date can't be trusted due to several unknowns, as young-Earth creationists, they all had different assumptions about the age of these samples. They thought they were relatively young and therefore had not gone through many half-lives. They expected to find C^{14} present. They were dated by state-of-the-art geochronology labs and all the samples tested (assumed to be tens to hundreds of millions of years old based on their position in the geologic column) had measurable amounts of C^{14}. They also dated some diamond samples assumed to be a billion years old by secular scientists. Even these diamonds had measurable amounts of C^{14}.[7] This is totally inexplicable, if they were as old as the geologic timescale demands. If a diamond is actually about a billion years old, the C^{14} should have gone through 174,520 half-lives. It should be emphasized here that there shouldn't be any measurable C^{14} in the coal or diamond even after twelve half-lives!

Uranium^{238} → Lead^{206}

The RATE team also had a number of granite samples from New Mexico dated using the U^{238} → Pb^{206} method.

---

7 Larry Vardiman et.al., "Summary evidence for a young earth from the RATE project," *Institute for Creation Research* (Dallas, Texas: ICR and Creation Research Society, 2000), 753–756, https://www.icr.org/i/pdf/technical/RATE2-Summary.pdf, (accessed April 6, 2023).

According to the measured ratio of U^{238} to Pb^{206}, the rock samples were 1.5 billion years old. The uranium and lead were contained within tiny zircon crystals embedded in the granite. But there was something else in the crystals that didn't make sense if it was really that old. It was helium. As mentioned earlier, uranium undergoes alpha decay (an alpha particle is composed of two protons and two neutrons). When these are ejected during radioactive decay, each alpha particle nabs a couple of electrons from the zircon crystal and becomes a harmless helium atom trapped in the zircon crystal matrix. Though much, much slower than helium diffusing out of a balloon, the helium atoms diffuse out of the zircon crystals and out of the rock entirely. Here lies a major conundrum: *there was still helium in the granite*. Based on the ratio of uranium to lead, it can be calculated how many alpha particles were released during the decay process and, consequently, how much helium would be produced. They were also able to measure the diffusion rate of helium at different temperatures. This was very revealing. Based on diffusion rates of helium from zircons, there shouldn't be any helium in the rock if the granite was indeed 1.5 billion years old. In fact 58% of the helium produced was still trapped in the crystals and about 6,000 years worth of helium had leaked out of the granite. If you assume the rock was 1.5 billion years old (based on uniform decay rates), then helium should be completely gone (based on uniform diffusion rates). However, since it's not gone, you have to assume it miraculously diffused 100,000 times slower than it actually diffuses. However, if you assume the rock is about 6,000 years old based on helium diffusion rates, then you have to assume the rock underwent accelerated radioactive decay 6,000 years ago.[8] The take home message is that you can't have it both ways. You can't believe both uniformitarian rates (uranium decay rates and helium diffusion rates) because they have extremely contradictory results. You can reject both but you can't believe both. These results don't necessarily prove young-Earth creation, but they do show that you can't trust uniformitarianism.

There are many other flavors of radiometric dating, but we need not go into all of them to grasp the take-home message. When geochronologists are attempting to discover the age of a particular rock, they must make three tenuous assumptions about the unobserved past. This puts the whole enterprise in the realm of forensic (historical) science—a type of science that, for obvious reasons, has a level of certainty much, much lower than can be obtained through empirical science. When we cross-check radiometric dates with rocks of known historical age and find that the radiometric date is ridiculously off the mark, that should give us pause about the reliability of geochronological results. The main difference between blind dating and radiometric dating is that the latter is far more blind. In blind dating, in the colloquial sense, the mystery of the unknown is resolved once the date has occurred. Those who practice radiometric dating are almost as clueless after the date as they were before.

---

8   Larry Vardiman et.al., "Summary evidence for a young earth from the RATE project," 739–742, https://www.icr.org/i/pdf/technical/RATE2-Summary.pdf (accessed April 6, 2023).

CHAPTER 4

Designer Genes—What's the Difference Between Microevolution and Macroevolution?

One of the worst problems in discussing creation and evolution is the misunderstandings arising from opponents using different definitions of key words. To avoid that problem I want to define my words clearly and carefully. Many creationists lose all credibility because they throw the baby out (certain truths of biological change) with the bathwater (the metanarrative of macroevolution like bacteria to butterflies or baleen whales). When they do this, meaningful dialogue ceases if a creationist happens to be talking to an informed evolutionist. Here is one example of wrongly stating the creationist position: "Creatures can vary within a species; they just can't produce new species." In this chapter, I hope to clarify the issue so that you will clearly understand why genetic variation can include speciation (one species becoming two or more species), and also why biological change has clear limits.

Although evolution simply means change, we can't take it that way anymore. If we did, young-Earth creationists would be evolutionists. This same problem arises in the climate change controversy. Taken in the correct sense, everyone believes in climate change; but in the way climate alarmists have defined that phrase, reasonable people get pigeon-holed as climate change deniers. Just like climate change, evolution has taken on a lot more baggage and it means far more than simply change. In short, it means that all biological diversity on Earth arose from one single-celled life form. If we study this topic carefully, there is a good deal of difference between the minor genetic variations (microevolution) that all creationists embrace and *large scale* macroevolutionary change. To many evolutionists, the degree of change we can see today and the macroevolutionary change over eons is not considered a difference in mechanism but rather a difference in the amount of time involved. How do you respond if an evolutionist says to you, "If you believe that a little bit of change can happen in a little bit of time, then why not a lot of change in a lot of time?" Now at first blush, that question may seem to be a stumper if we grant him millions of years. You might refuse him millions of years for biblical reasons, but the lack of millions of years is not the real problem interfering with biological change. Even if we grant him deep time, hundreds of millions of years is not a biological blank check to generate unlimited change toward increased complexity. If you say to yourself, "He might be right, if life on earth has a ton of time to evolve," you missed the point. Even if you grant him billions of years, biological change would still have its limits due to lack of intelligent input. It is of the utmost importance for Christians to understand where these limits are and what the mechanisms of change can and can't do.

Now let's get down to definitions. First, let me define certain mechanisms of change that creationists and evolutionists can agree on.

MICROEVOLUTION

Microevolution is a change in gene frequencies in a population over time. Some texts define *evolution* this way, which really blurs the distinction between fact and fiction. Now let's tease apart this definition to see if this is actually a true phenomenon. What are genes anyway? In Chapter 7, we will find that many genes or sets of genes code for RNAs, which may code for proteins,[9] which in turn may directly or indirectly affect the structure, physiology, and/or behavior of an individual organism. Many genes code for RNAs that regulate the expression of genes that code for proteins. So when you talk about genes you are talking about the hidden genetic cause of much of an individual's structural, physiological, or behavioral traits. There is also an increasing awareness that epigenetics[10] is extremely important in an individual's structural, physiological, and behavioral traits. The word *frequency* refers to the proportion of certain genes present in a population. What is a

---

9   Many RNAs have other functions and don't code for protein

10   The study of changes in organisms caused by modification of gene expression rather than alteration of the genetic code itself. Definition from Oxford Languages dictionary.

population? A population is a group of individuals of the same species living in a specifically defined area. Here's a silly scenario. Let's suppose 100 Sudanese move to a small town in Minnesota that is composed of 9,900 fair-skinned Scandinavians. Let's also suppose that none of the Scandinavians have any genes coding for dark skin. And the Sudanese, similarly, have no genes coding for fair skin. The total population is 10,000. One hundred people out of 10,000 carry genes coding for dark skin, so the frequency for those genes is 0.01 or, more simply put, 1%. If for whatever reason that frequency changes over time, it is called *microevolution* (I will use the term *genetic variation* interchangeably). Below is a list of factors that can change gene frequency.

1. Gene flow—the Scandinavians and the Sudanese intermarry and have kids causing the dark skin genes to become more widespread in the growing population.

2. Immigration—people of either ethnic group move to this town.

3. Emigration—people of either ethnic group or their offspring leave this town.

4. Genetic recombination—this is the normal cutting and shuffling of the "deck" of genes prior to them being dealt out to sperm and egg cells. When fertilization occurs, a new "deck" of genes is generated. No gene is new, but new combinations of genes are produced.

5. Mutation—a mutation changes the melanin gene causing the melanin to be a slightly different color,[11] or it changes how much melanin is produced.

6. Natural selection—Suppose increased solar flares and climate change (sun screen is outlawed) result in many clear days in Minnesota causing a higher death rate due to skin cancer in the fair-skinned portion of the population. The frequency of dark skin genes increases.

7. Genetic drift—Some environmental disaster. Minnesota gets their first volcano, and its eruption indiscriminately destroys half the town with a lava flow killing 5,000 people which happened to contain all 100 Sudanese. The frequency of fair skin genes increases.

Although the above scenarios are contrived and silly, there is nothing silly about the outcome if those hypothetical events all occurred. These kinds of mechanisms are things we can all understand and observe in nature, and none of them poses any threat to the belief in creation.

Thinking it Through Critically

Many creationists don't like to use the word *microevolution* because it often causes confusion and lures impressionable

---

11 Skin color is the trait so there is really no novel gene produced but rather a *new version* of a pre-existing gene called an *allele*. Throwing a new version of the skin color gene into the mix shifts all gene frequencies.

students into the belief of *macroevolution* by blurring the distinctions between the two (there is also only one letter difference between the two words but worlds of difference in content). It would be better to call it *genetic variation*. Regardless of the term used, it is important to carefully describe what is truly occurring in genetic variation (or microevolution) so that we can easily distinguish it from the unsubstantiated science fiction of *macroevolution* with its presupposed unlimited capacity for change.

Let's briefly examine the nature of the aforementioned factors contributing to microevolution.

Immigration and emigration are tied to gene flow. Migration simply shuffles variant genes around geographically between different populations of the same basic *created kind* but it doesn't add or subtract from the gene pool of that basic kind. There is nothing truly novel that bestows some physical feature previously nonexistent—no magical appearances of new information gained. You might ask, "Isn't dark skin or kinky black hair novel in this pure Scandinavian population?" Yes, but genes that code for skin color (or hair color and texture) are not. A gene that adds a *new version* to an already existing set of genes does not contribute to the process of macroevolution. The successful hybridization and the production of offspring after migration, is called **gene flow**. This introduces new versions of genes (of already existing traits) into the gene pool. Another example of this is when the town mutt jumps the kennel fence of a champion poodle and shortly thereafter the breeder, to his dismay, finds an unexpected litter of puppies. New versions of dog genes were added to this poodle population but no new trait was added to the dog *kind*. Again, nothing completely new is added to the *created kind*.

Genetic recombination is a wonderful biological mechanism to produce amazing diversity from within the unity of a kind. Basically, gene "decks" of both parents are shuffled, cut, and dealt out to sperm or eggs (during meiosis). Considering the number of genes, the possible combinations are astronomical, which is why siblings are always different (except identical twins) yet from the same two parents. Again, there is no brand new genetic information added. Your little brother may be different, but he doesn't have feathers or two horns growing out of his head.

Mutation is the only mechanism that produces anything "new." But actually it really isn't totally new. It is an old gene that gets altered into a new version of the same basic gene. Different versions of the same gene, as previously mentioned, are called alleles. Mutations at best produce new alleles that increase the number of options for the same trait (for example, eye color, hair color, and skin color have many alleles). Some of these alleles may have been built into the original creation but others may have been the result of mutations. Mutations are a mistake in the sequence of a gene. This usually alters the amino acid sequence of the protein being made. This gene directly or indirectly affects some particular trait, either all by itself or in concert with other genes. Most mutations are either neutral or harmful,[12] rarely beneficial. But even when we closely examine beneficial mutations, no increase in complexity was added. In fact, beneficial mutations are often

---

12 Even the neutral mutations are slightly harmful but are so slight that they are not selected against.

from a loss of genetic information or function. Two examples have been touted as excellent evolutionary mutations. The first is penicillin resistance in bacteria. In reality, the mutation did not generate anything novel. The bacteria already had the gene that codes for the enzyme *penicillinase* (which breaks down penicillin). However, most bacteria do not produce enough of it to effectively deal with medically prescribed doses. When certain disease-causing bacteria have a mutation in the genetic switch (which is supposed to keep penicillinase production at moderate levels), all of a sudden there is no restraining mechanism and penicillinase is cranked out with abandon. This broken gene switch, of course, is only an advantage to mutant bacteria if they are exposed to high levels of penicillin. In competing with the non-mutant bacterial strains in a normal environment, the mutant strain is quickly eliminated because it is spending way too much of its food and energy making inordinate amounts of this enzyme, which slows down its reproductive rate. But biology professors usually don't want to bog down introductory biology students with all these details, so the student walks away thinking that when some organism encounters a hardship it evolves some totally new innovation to deal with it.

Insecticide resistance is also heralded as a great example of evolution. In agricultural ecosystems, crops are regularly sprayed with insecticide to keep certain pests at bay. Frequently, certain individuals will get a mutation that renders them resistant to the poison. This is another example of nothing novel. The insects already had the genes coding for enzymes to detoxify certain families of harmful chemicals. Mutations broke a switch (like the bacteria example) so that it could crank out enough enzymes to cope with the high doses of insecticide. Another way crop pests have thwarted our efforts is by mutations altering the shape of the actual enzyme so that it becomes more effective in inactivating or detoxifying the poison. So once again, no new information was generated from scratch.

Natural selection is a conservative process among living creatures. Evolutionists make it sound as if mutation coupled with natural selection has creative power. Not so. People who are well-informed on natural selection know that it simply weeds out individuals (possessing traits and the genes encoding those traits) that are ill-suited for a particular set of environmental conditions. My previous example of fair-skinned Scandinavians having higher mortality due to skin cancers is a hypothetical example of natural selection. If the solar radiation increased (assuming no one used sunscreen and all wore sleeveless shirts), fair-skinned people (bearing fair-skin genes) would be gradually removed from the population due to a higher incidence of fatal skin cancers or emigration to a cloudier country. As a result, the darker-skinned people (bearing dark-skin genes and being more suited to the sunnier environment) thrived and had children who inherited those genes and the dark skins they produced. This is why it is called natural selection. Nature "selects"[13] those traits that are suited for

---

13 It has been noted that the word *selection* implies purpose or direction toward a goal. It must be understood that there is no purpose or goal in evolution (which evolutionists acknowledge). Therefore, to compare natural selection to artificial selection is not accurate because the latter includes the goal or purpose of the breeder. Not so the former. Nevertheless, even if we grant purpose or goal in natural selection, it still would just be as limited as artificial selection. Breeders don't write new genetic information; they only select from what is there. The same is true for so-called natural selection.

the environment and eliminates those traits that aren't. The stringency of natural selection is called *selection pressure*. In the wilderness, selection pressure is more intense. Due to human medicine, technological innovations, and compassion, we have sheltered ourselves and our loved ones from the intense selective pressures experienced by animals and plants in the wild. But the bottom line of natural selection is that it doesn't generate any new genetic information; it just prunes or weeds out any living being that isn't suited for that environment (along with the genes that made it unsuitable). This will change the genetic profile of a population and therefore conforms to the ideas of genetic variation or microevolution; but it is a far cry from macroevolution.

Genetic drift differs from natural selection in that certain individuals are removed from the population due to bad luck or ill fate. Their elimination wasn't because they had ill-suited genes (and consequently ill-suited traits) for a given environment. Instead some disaster indiscriminately took out a substantial chunk of a small population which may have drastically reduced or eliminated a certain allele (and its corresponding trait) from the population. It is only similar to natural selection in that the genetic profile of the population changes due to the removal of certain individuals.

There are many other interesting facts about the above mechanisms that can be gathered from a more in-depth biology text. My only purpose here is to present a brief sketch of the major evolutionary mechanisms and demonstrate why they cannot generate new genetic information bestowing novel traits on organisms whose ancestors never had those traits or the genes to code for them. In all of the above examples, it was a rearrangement, redistribution, removal, or remodeling of the existing genetic information.

MACROEVOLUTION

Most textbooks don't give a crisp definition to this concept. I won't speculate too much on the reasons why, but I do think that they blur the distinction (either deliberately or inadvertently) between microevolution and macroevolution so that students will attribute the wholesale changes of macroevolution to an accumulation of minor genetic changes already discussed. Whatever their reasons, it works. Most students wind up thinking that all the mechanisms contributing to microevolution can, over the long haul of millions or billions of years, gradually accumulate enough small changes that result in a host of big changes, e.g., fish to amphibian, amphibian to reptile, reptile to bird, etc. Since I don't like their definitions, I will propose one that makes a clear distinction between the fact of microevolution (genetic variation) and the fiction of macroevolution.

Macroevolution—the addition of new or novel traits (neomorphs) within a population when the ancestors neither had those traits nor the genetic information to code for them.

To reiterate my summary of the list of factors contributing to genetic variation, I wrote, "In all of the above examples, it was a rearrangement, redistribution, removal, or remodeling of the existing genetic information." In stark contrast, macroevolution requires **"additional traits & genetic information encoding those traits"** to generate the kinds of changes that need to occur between bacteria, beetles, bass, bats, and baleen whales. Their brainwashing success is due to slowly sprinkling out all these additions over such a vast amount of time and directing our attention to all the similarities that these disparate life forms share. They avoid emphasizing the huge differences in form and information content (that may generate difficult questions).

NEOMORPHS: NEW ADDITIONAL TRAITS (IF COMPLEX LIFE EVOLVED FROM SIMPLER LIFE)

Even a quick survey of life would generate a countless array of creatures possessing **neomorphs. Neomorphs are structures or traits that their presumed ancestors did not have (nor the genetic information to code for them).** Many of these additional traits would be subtle physiological or behavioral traits so non-biologists would have difficulty seeing anything tangibly novel when comparing them to their ancestors. So I will restrict my discussion to a handful of clear-cut morphological traits that presumably had to evolve from critters that lacked those traits.

The Feather

You don't have to be a biologist to know the most obvious feature of birds. Yes, the feather. If you follow the normal evolutionary progression, which is common knowledge, birds evolved from some featherless reptile. Even though there is evidence that certain dinosaurs had feathers, there still needs to be some ancestral featherless dinosaur that gave rise to feathered dinosaurs or birds or both. My point is that at some point feathers had to evolve…from scratch. Feathers are not elaborate modified scales. It was previously thought that genes coding for scales were mutated and eventually gave rise to genes coding for feathers. Evolutionists now know this is not the case. In modern birds, the genes that code for scales *are not* the genes that code for feathers. These two structures have their own genetic information encoding them. Is there any empirical evidence that unguided mutation can transform spare copies of scale genes (or any spare DNA for that matter) into feather genes? No. If you look at the exquisite microscopic structure of the avian feather (and how it develops) it becomes apparent that this is beyond the scope of unguided mutation. The avian feather consists of a central shaft. Arising from this on both sides are hundreds to thousands of barbs (see Figure 5).

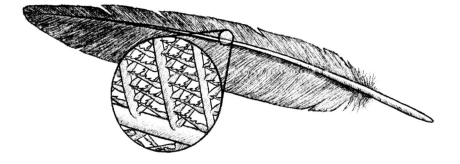

Figure 5: A feather.

The barbs interlock to form the vane. This is made possible by thousands of barbules that arise from all the barbs. Certain barbules have an amazing array of hooks that interlock with the barbules of an adjacent barb. If you think that undirected mutations acting on spare DNA can encode the architecture and orchestrate the construction of the feather, you lack an intuitive understanding of probability. That would be analogous to thinking a toddler with a pen and a big blank sheet of paper could draw up exact architectural blueprints of a Gothic cathedral with detailed building instructions. The blueprints are one thing, but the information to direct and coordinate the actual builders at the construction site so that the cathedral is successfully built in proper sequence, is quite another. A materialistic skeptic egghead probably wants the respective probabilities of feather DNA arising from mutations and cathedral blueprints arising from a scribbling toddler so that he can make an informed decision. I don't have the numbers but even if I did, this kind of egghead would conclude that the scribbling toddler-cathedral event would never happen (due to the astronomically low probability) but the mutation-feather would (never mind the probability). If the presence of cathedral blueprints needs to be explained, they can simply be attributed to a highly skilled architect over a scribbling toddler. That's considered scientifically reasonable because a sufficient cause for cathedral architecture happens to be an observable person. But it is *not* considered scientifically reasonable even when the only sufficient cause for the feather's architecture is an omnipotent Creator. Why? Anything that transcends natural processes is ruled out a priori. In our academic hubris, we decide that unintelligent, physical causes are the only causes we are allowed to consider, *even if they are grossly inadequate.*

The Humble Turtle Shell

Another obvious neomorph is the turtle shell. In evolutionary thinking, it had to evolve from a shell-less lizard-like reptile. It is always preferable to the evolutionist for some new innovation like the feather or the turtle shell to be some modification of a pre-existing structure. In other words, neomorphs are most problematic for them. They usually attempt to explain their appearance as modifications of pre-existing structures. For the feather, it was wrongly presumed to be a modified scale and for the turtle shell, it was wrongly presumed that the upper shell was a modified rib cage (ribs fused to each other and to the vertebral column) and the lower shell was a modified (enlarged) sternum. But this is a gross oversimplification. It is now known that the upper shell (the carapace) is not simply an enlarged and fused rib cage and backbone. There is additional structural and regulatory genetic information required to turn the dermis into bony material and to fuse the whole works: the ribs, the vertebrae, and the ossifying patches of dermis (osteoderms) to each other and to themselves. All this must also be overlaid with and fused to a hardened fingernail-like keratinized epidermis (containing the color pattern of the turtle) and still house and protect a mobile creature within. This is the only vertebrate that has its pectoral and pelvic girdles inside the rib cage. Go figure. The bottom shell (the plastron) is not just an outsized sternum. The osteoderms form on the turtle's belly and fuse with the sternum to form the plastron. It is also overlaid with the hardened skin. The plastron, on the two sides just behind the forelegs, fuses with the upper shell to form the bridge. By the way, most land vertebrates, reptiles (other than turtles), and mammals ventilate their lungs by moving the rib cage like a bellows. Of course turtles can't do this, so they have other ways to ventilate their lungs, i.e., the movement of their limbs. An evolutionist must explain the evolution (from scratch) of the carapace and plastron but also this ventilating mechanism custom-made for turtles. There are other structural subtleties that would have to evolve from scratch that set turtles apart from their non-turtle ancestors, but I want to restrict my discussion to just a few examples to make the point. These obvious and subtle differences between turtles and their presumed reptilian ancestors involve *additional* genetic information to encode the stuff and its assembly instructions. The turtle body plan was designed. It wasn't an unimaginably long series of lucky structural and developmental mutations on a lizard-like creature. The burden of proof is on the person who thinks otherwise. Why? His position flies in the face of all empirical evidence substantiating Information Theory (which maintains that only intelligence can produce specified information, including genetic information). He must empirically show that highly adaptive novel structures (like feathers and turtle shells) can appear through unguided mutations. If experimental time frames prohibit a successful demonstration of turtles evolving from mutated lizard-like reptiles or other such unlikelihoods, theoretical scenarios must factor in the infinitesimally low probabilities (no matter how painful).

The Ptilinum

My first two examples (feathers and turtle shells) were structures that everybody is familiar with, but keep in mind that any biologist who has a decent grasp of the diversity of life would be able to rattle on all day listing discrete anatomical structures of creatures whose ancestors lacked them entirely. Spreading the evolution of these structures over eons of time does not eliminate the basic problem of getting something from nothing. My next example is probably unknown to most folks except entomologists. It is one of my favorites because it seems to evoke the strongest 'gee whiz' responses. This is simply because it's a bizarre structure performing a bizarre task and is outside of people's common knowledge and experience. Biological structures we see daily in ourselves or our pets are just as complex and reveal God's glory just as powerfully (if we take the time to examine them); but since they are so common and familiar, we take them for granted and don't think they are as problematic to evolution as the bizarre examples. Unfortunately, many entomologists still fail to recognize the Architect of the creatures they esteem. I would like to present several insect structures (and behavior) that work together to perform a certain task, which when viewed by those who acknowledge God's glory will clearly expose the insanity of believing in the creative powers of mutation and natural selection.

All flies (Order Diptera) in the group known as Schizophora (which includes the housefly and bluebottles) have a small, short seam on the front of their face arching over the base of their antennae and between their large compound eyes (Figure 6). At first, this seam seems to be an insignificant crack on the fly's face. In actuality, it marks the edge of a door that popped open during the fly's emergence from its puparium. As many of you know, the larval fly is known as a maggot.

Figure 6: A Schizophoran fly face.

During metamorphosis, the pupa stage is encased in the last larval (maggot) skin. This skin hardens into a thin leathery cocoon-like container called a *puparium*. During the emergence of the adult fly, not only does it have to crawl out of its pupal skin, but it also must exit the outer puparium (yes, kind of like Matryoshka dolls). The puparium is roughly the size and shape of a grain of puffed rice (Figure 7).

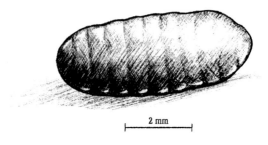

Figure 7: A puparium.

One end has a circular seam forming a cap which pops open under internal pressure. The fly, after emerging from its pupal skin, cannot simply nudge this cap open. So at this point, certain muscles contract surrounding the body cavity containing the blood-like fluid called *hemolymph*. This contraction propels the hemolymph forward into the head (talk about a head rush). The *ptilinum*, a balloon-like sac within the head, inflates. This increase in pressure causes the door on the fly's face to open like a drawbridge. The ptilinum inflates quickly with hemolymph and bulges out of the opening like a vehicle's air bag. The inflating ptilinum exerts pressure on the inside of the puparium's cap, causing it to pop open. The adult fly emerges, the ptilinum deflates and is withdrawn into the head, and the door closes back over the opening, never to open again.

There are several amazing, integrated traits worth noting:

1. the ptilinum within the head,

2. the door on the fly's face,

3. the circular seam on the puparium forming a cap,

4. the muscular contraction (at an appropriate time) which forces the hemolymph into the ptilinum causing the door to open and the puparium cap to open,

5. the permanent repackaging of the ptilinum into the fly's head.

There are other anatomical and physiological subtleties that make this process even more amazing, but I will stop here lest you begin feeling guilty destroying this amazing biological machinery using your flyswatter, and I certainly don't want you to refrain from that. According to evolution, Schizophora flies had to evolve from a simpler group of flies (the Aschiza) not possessing these gadgets nor the genetic information encoding them. Keep in mind that if macroevolution is true, traits 1–5 (and others not mentioned) affecting the emergence of a housefly had to arise from unguided mutation.

Have experimental mutations in any creature ever modified or produced an array of traits so exquisitely integrated and coordinated to accomplish similar feats? No. This is but one of countless biological phenomena crafted by a mighty Creator Who cannot be replaced by non-creative, non-intelligent mechanisms of mutation and natural selection. Rather both mutation and natural selection are ordained by God as mechanisms to allow creatures to adapt to a changing environment.

The same impasse outlined in Chapter 8: Micro-machines is the same here. Evolution of irreducibly complex molecular systems requires genetic information encoding, directly or indirectly, the components of the system. However, in subcellular systems it is easier to point to discrete genes coding for particular proteins. Although tissue-organ level neomorphs (feathers, turtle shells, ptilina and accessories) are even more complex, the genetic information responsible for them is not thoroughly sequenced or understood. Nevertheless, not being able to call the entire system irreducibly complex does not detract from the incredible complexity but rather underscores it. A car, though obviously designed, is not irreducibly complex. You can knock off the hubcaps, the bumper, the hood ornament, etc., and it still works.

INFORMATION THEORY IMPASSE

Can genetic recombination, mutation, natural selection, genetic drift, and gene flow generate designer genes that code for neomorphs (new or novel structures) in creatures when the ancestors did not have those structures or the genes to code for them? No. Why not? Again, all this new stuff requires new genetic information. Old "spare" genetic information that was formerly coding for something else has never been shown to mutate out of an old function to gain a totally new and useful function. Empirically, it hasn't been done, and according to Information Theory it cannot be done. Neomorphs require genetic information, and information requires intelligence. Genetic information coding for exceedingly complex biological structures and systems requires intelligence of the highest order. If it is intelligence far beyond the capacity of the greatest human engineer, how can the scientific community say with impunity that it wasn't designed? If the creative Intelligence is beyond the physical realm, and if they can't see it, then it must not be. That kind of hubris takes the breath away.

THE CREATIONIST EXPLANATION OF NEOMORPHS

> So God created the great sea creatures and every living creature that moves, with which the waters swarm, according to their kinds, and every winged bird according to its kind. And God saw that it was good (Genesis 1:21).

And God made the beasts of the earth according to their kinds and the livestock according to their kinds, and everything that creeps on the ground according to its kind. And God saw that it was good (Genesis 1:25).

Neomorphs never arose through evolutionary processes. At the creation, neomorphs were never added as an afterthought of God or produced through unintelligent factors of genetic variation (mutation, natural selection, etc.). All birds were endowed with feathers (along with the genetic information necessary for their development) according to their kinds. Turtles were endowed with turtle shells (along with the genetic information necessary for their development) according to their kinds. Schizophora flies were endowed with ptilina and accessories (along with the genetic information necessary for their development) according to their kinds.

ANALOGIES TO MAKE IT STICK

I want to leave you with a few analogies to illustrate why the factors of genetic variation are incapable of generating neomorphs. Remember, genetic recombination, genetic drift, gene flow, mutation, and natural selection can, at best, rearrange, redistribute, or remodel existing genetic information and, at worst, damage or remove genetic information. It cannot develop or design new specified genetic information to produce new structures or new functions. Only intelligence can do that. Am I repeating myself? Good, because it's important to get this.

If you owned an old computer and you wished it would do a lot more, store a lot more, and work a lot faster, what would you do? Would you rub a magnet on its hard drive and hope that your wishes would come true? No, of course not. If you had the money, you would purchase the latest computer loaded with all the whippy-doo software you're coveting (or download it onto your current computer). How was all this software produced? Was it produced by a binary random number generator that generated a lot of 1s and 0s that were burned onto a CD by chance? How would that CD sell in the IT marketplace (market pressure is analogous to natural selection)? Not well. Why? Ones and 0s don't produce useful information (or any information) apart from intelligence. If it doesn't sell, then it no longer is reproduced and thankfully fails miserably (selected against). Software that survives the market is somehow appealing and beneficial to computer users. And guess what…the kind of software that sells was generated by intelligent and clever software engineers *that know what they're doing and know what you want your computer to do!* If it is a product that satisfies a market need (i.e., pressure), it is purchased and consequently reproduced according to demand (selected).

If a university received funding to build a molecular biology addition to the science building, are random errors introduced into the digital blueprints of the original science building, printed out, and given to the contractor? Hopefully not. But if they decided to build it according to the "mutated" blueprints it would be a faulty replica of the first, not a new building serving the current needs of the institution. If you want a unique addition with molecular biology in mind, you must have intelligent and skilled architects draw up new plans with the new purpose in mind. I'm not an architect, but I'm sure it requires a lot of thinking and planning, along with a lot of architectural know-how, if it is to be highly useful to all the molecular biologists who will eventually be working in it.

Similarly, in the automotive industry, driver-side airbags did not come about from the accidental changes in the car's design blueprints. Airbags are obviously designed. Once Research and Development is satisfied with the design of the product, it will be manufactured, tested for safety, and eventually put on the market. Obviously, airbags have a selective advantage over airbag-less cars. I don't dispute that. We are all well aware that the new product has a selective advantage over the earlier model such that the new often makes the old go extinct. But that fact does not explain the product's origin. The late Ernst Mayr (the Jedi master of evolution during the twentieth century), in his book *What Evolution Is*, uses the evolution of cars as an analogy for macroevolution. I'm glad he did because it reveals the fundamental flaw in evolutionary thinking. Unfortunately, Darwinists are such uncritical thinkers (when it comes to defending Darwinism) that they don't see the flaw. He writes,

> It [evolution] was as progressive as the development of the modern motor car from such early types as Ford's Model T. Each year the manufacturers of motor cars *adopted new innovations* and these were then exposed to the selective pressure of the market. Many models with certain innovations were eliminated; the successful ones formed the basis for the next level of innovation. As a result, the cars improve from year to year, becoming safer, faster, more durable, and

more economical. Surely the modern car represents progress (emphasis mine).[14]

He grossly overlooks the many sources of automotive intelligent design to emphasize the power of "the selective pressure of the market." But we can't overlook the genius of these automotive engineers. Every successful automotive innovation was the work of bright teleologically-driven engineers (that is, purposeful), and we are all grateful for them. Mayr knows that because he goes on to say, "However, Darwinian progress is never teleological." Well, his analogy sure is.

The biological world is filled with creatures possessing anatomical structures which were not possessed by their supposed ancestors. The design features and the development of these neomorphic structures far exceed the complexity of any automotive innovation since the Model T. What was the source of the innovation: gene flow, genetic drift, genetic recombination, mutation, natural selection, or all of the above? All these mechanisms fall miserably short of explaining complex biological innovations accruing between bacteria and baleen whales. But you'll have to conclude wrongly if you must exclude God.

If inferring a Designer/Creator is so obvious, why are so many people committed to believing in such a counterintuitive explanation? Although there are many answers to that question, I would like to focus on a few reasons why I think so many are committed to and convinced of Darwinism:

---

14 Ernst Mayr, "Macroevolution," essay in *What Evolution Is* (New York, NY: Basic Books, 2001), 215.

1. Public education is committed to keeping God out of any causal role in producing the presence, complexity, and diversity of biological life.

2. Public education presents strong evidence for microevolution/genetic variation (in persuasive and scientific language with no alternative offered) and then blurs the distinction between microevolution and macroevolution.

3. Most students don't ask whether there is an actual distinction between them. Even the smart ones.

4. Many students are predisposed to accept any proposition suggesting that God is either nonexistent or irrelevant in matters pertaining to the physical universe (including their lifestyle). Simply put, they want a world without God; no One to obey; no One to praise or worship. If Darwinism claims to produce all biological complexity including their brain and body, then there is no Supreme Being to Whom they must give an account for their actions. Therefore they are only accountable to themselves (and other humans). It claims to explain why "guilt" is merely a neurochemical reaction in the brain (a product of evolution) rather than a spiritual phenomenon that occurs when we transgress the law of God.

CHAPTER 5

Testimony of the Entombed—Does the Fossil Record Reveal Common Ancestry?

Despite the many pieces of evidence supporting creation and intelligent design, it is widely assumed that the fossil evidence is the pièce de résistance of evolutionary theory. Actually, over one hundred sixty years ago, Darwin was acutely aware that the fossil record wasn't all that helpful. This quote from *The Origin of Species* sums up his assessment of the fossil evidence (at that time) supporting his theory of gradualistic evolution:

> Why then is not every geological formation and every stratum full of such intermediate links? Geology assuredly does not reveal any such finely graduated organic chain; and this, perhaps, is the most obvious and serious objection which can be urged against the theory. The explanation lies, as I believe, in the extreme imperfection of the geological record.[15]

---

15 Charles Darwin, *The Origin of Species: 150th Anniversary Edition* (National Geographic Books, 2003).

Darwin knew that paleontology was a relatively young science, so he was hopeful that future fossil finds would unearth many intermediate links so desperately needed for his theory. He didn't expect all intermediate links to be found because he knew that many fossils would have been destroyed by geological activity over vast amounts of time. That is what he refers to as *the imperfection of the geological record*. Shortly after the publication of *The Origin of Species*, *Archaeopteryx* (an extinct bird genus with several dinosaurian characteristics), was discovered and celebrated as one such intermediate link (this strange bird doesn't quite fit the bill [no pun intended] but more on that later). Other "intermediate links" have been found between different forms but they are the exception, not the rule, and are easily explained by the creation model. Though countless fossils and many more species have been unearthed since Darwin, the gaps have been accentuated rather than filled. In short, a closer look at the fossil evidence reveals another Achilles' heel of evolutionary theory. However, we need good spectacles to view the evidence. In this chapter we will do just that. My objective in this chapter is to see if the fossil record is in line with gradualistic evolution or not. Is there one great evolutionary tree having one trunk gradually forming several to dozens of major branches of life (kingdoms, phyla, classes, orders, etc.), each in turn, gradually diversifying into hundreds of medium-sized twigs (families and genera) and thousands to millions of small leaves (species)? Or will we see fossil forms distinctly different from other fossil forms from the earliest rocks until now, with a curious absence of intermediate forms connecting them to each other and to a common ancestor? If all life arose from one unicellular life form, you would expect to see a continuum of life forms in the fossil record—twigs joining twigs to form branches, branches joining branches to form the trunk, and so on. But before we look at what the fossil record reveals, I would first like to briefly discuss the conditions under which fossils are generally formed.

Fossils form under extraordinary conditions, conditions you don't see every day. Suppose you were driving on a two-lane country road. In the middle of your lane is a road-killed opossum. It could be a driving hazard so you dutifully pull over, get out, grab it by the tail, and heave it into the muddy creek in the ditch next to the road. It sinks down into the murk. Having done your good deed for the day, you get into your car and cruise down the road. Will that opossum become a fossil? No. Why not? Even though it was quickly covered by a layer of silt and mud, it won't fossilize simply because the conditions are not extreme enough. It won't be buried fast or deep enough. Soon all sorts of aquatic insects and other invertebrates will feast upon its carcass. In addition, a whole host of microorganisms (bacteria, fungi, water molds, and other protozoans) will pile on to assist its going the way of all flesh. They will handily decompose anything the scavengers miss. In short, everything will be consumed...including the bones. In order to fossilize, there must be a tremendous amount of eroded sediment like silt, mud, sand, etc., that is carried by wind or water. Plants or animals must be *rapidly and deeply* entombed in the sediment, so much so that some of their remains (usually hard structures like bones, shells, exoskeletons, etc.[16]) can undergo the mineralization[17] process before every body part is consumed by aquatic scavengers and decomposers. Usually soft tissues like muscles, internal organs, and skin don't last very long even when buried deeply. The next thing that needs to happen is that the sediments need to become sedimentary rock. For this to happen, the sediment needs to have cementing agents (such as calcium carbonate, also known as lime[18]) present in the deposit while it dries out. In fact, artificial sedimentary rock is made when concrete is poured and cured. The lime causes the sedimentary particles (in this case, sand grains and pebbles) to rigidly lock together as it dries out, causing the sediment to become sedimentary rock (concrete).

TWO GEOLOGIC PARADIGMS

The paradigm that most secular geologists use to interpret the rocks is *uniformitarianism*. This term was coined by William Whewell when reviewing Charles Lyell's *Principles of Geology* (1830–1833). This paradigm operates under the assumption that rocks, fossils, and formations were slowly formed over enormous amounts of time by slow, uniform, and continuous working processes that can be observed today. It has been summarized by the oft-quoted phrase,

---

16 These are more likely to fossilize because they are less likely to be quickly consumed by scavengers and decomposers.

17 The process by which organic material of an organism's remains is replaced by inorganic chemicals or minerals.

18 This is a common example of a cementing agent, but there are others.

"the present is the key to the past." In other words, in order to figure out *what* shaped the earth and *how long* it took in the distant past, you must study the processes at work today. For example, if sediment is laid down in a delta at a rate of millimeters per year under normal conditions, that rate is what you use to determine how long it took to lay down thousands of feet of sedimentary rock. Although secular geologists recognize that catastrophic events have periodically interrupted the prevailing uniformitarian conditions, they are considered minor factors in the overall earth-shaping processes throughout Earth history.

The other paradigm widely held before *Principles of Geology* was published was *catastrophism*. This paradigm operates under the assumption that rocks, fossils, and formations were rapidly formed due to catastrophic events in a relatively short period of time. This paradigm is generally held by biblical young-Earth creationists, who assume that Noah's Flood was *the* catastrophic event responsible for most of the sedimentary (and igneous) formations covering the earth's crust. Just as those who hold to uniformitarianism acknowledge catastrophic events, catastrophists also acknowledge uniformitarian processes at work now and for long periods (hundreds of years) before and after the Flood. Even though our view of Earth history is mostly uniformitarian, the reason we aren't uniformitarians is because we believe most of our visible geology is due to catastrophic events. Since the earth is a little more than 6,000 years old, uniformitarian processes don't accomplish much in that amount of time.

With that as a backdrop, let's look at the fossil record. It is not just creationists who see a pattern that is not consistent with uniformitarianism or gradualistic evolution. The late Stephen J. Gould, arguably the greatest evolutionary paleontologist of the last half of the twentieth century, was shockingly honest about the overall pattern of fossils. He writes,

> The history of most fossil species includes two features particularly inconsistent with gradualism: 1. Stasis. Most species exhibit no directional change during their tenure on earth. They appear in the fossil record looking much the same as when they disappear; morphological change is usually limited and directionless. 2. Sudden appearance. In any local area, a species does not arise gradually by the steady transformation of its ancestors; it appears all at once and 'fully formed.'[19]

Niles Eldridge, another paleontologist, concurs:

> He [Darwin] prophesied that future generations of paleontologists would fill in these gaps by diligent search…It has become abundantly clear that the fossil record will not confirm this part of Darwin's predictions. Nor is the problem a miserably poor record. The fossil record simply shows that this prediction was wrong.[20]

Many more quotes from numerous scientists making similar claims have been compiled at the website just footnoted in the previous two quotes. Apparently the pattern hasn't

---

19 Richard William Nelson, "Fossil Record By the Decade," Darwin, Then and Now, October 4, 2014, https://darwinthenandnow.com/glossary/fossil-record-by-the-decade/.

20 Richard William Nelson, "Fossil Record By the Decade," Darwin, Then and Now, October 4, 2014. https://darwinthenandnow.com/glossary/fossil-record-by-the-decade/.

changed much in the last 160 years. Darwin's hopes have not been fulfilled; they've been dashed. But let's see for ourselves. Let's look at the pattern that Gould and these other scientists have seen and commented on that is not consistent with Darwin's gradualistic evolution.

FROM UNICELLULAR TO MULTICELLULAR LIFE

In order for you to see the pattern of fossil evidence, I will present a series of diagrams showing the ranges of various fossil groups—sometimes phyla, sometimes orders, sometimes families, etc. I will start at the bottom of fossil bearing rocks. According to the geologic timescale, an enormous period of time collectively called the Precambrian spanned from when the earth formed (supposedly 4.5 billion years ago) to the beginning of the Cambrian Period. Most experts believe that the first unicellular life arose from the primordial soup about 3.7 billion years ago. They also think that unicellular life reigned from then to about 600 million years ago, at which point in time they think that multicellular life began to evolve. We will discuss the impossibility of unicellular life self-assembling from simple organic building blocks in greater detail in Chapter 7, but for now suffice it to say that to go from unicellularity to multicellularity is on par with the plausibility of a fat man jumping to the moon. The idea of the spontaneous infusion of highly specified genetic information into a unicellular life form, that orchestrates and coordinates the development of a *multicellular* life form that actually works (with all its integrated tissues, organs, and systems), is laughable. Again, mutation doesn't generate novel specified information. Natural selection only selects existing variants; it can't innovate anything. But back to the rocks. The first geologic period that is burgeoning with fossils is the famous Cambrian Period (from 541 million to 488 million years ago). What is so remarkable about it is that so many disparate animal phyla suddenly appear entombed in these rocks with no evolutionary ancestors in the Precambrian rocks. Yes, there are some trace fossils (worm burrows and tracks) which indicate some animal life, but these offer no evidence of evolutionary antecedents of the Cambrian phyla which can be finally traced back to one universal common ancestor deep in the Precambrian Eon. There is also a strange group called the Ediacaran fauna. Based on a detailed analysis of their structure, these also aren't good candidates to be evolutionary ancestors to any animal phyla in the Cambrian Period. This sudden appearance of animal life is called the *Cambrian Explosion*. Some of the major phyla include Porifera (the sponges), Cnidaria (the sea anemones, coral, jellyfish, etc.), Annelida (the segmented worms), Mollusca, Arthropoda (mostly crustaceans and trilobites), Echinodermata (the sea stars, sea urchins, sea lilies, etc.), Chordata (cephalochordate fossils), and over a dozen more small, obscure phyla. This array of different phyla (complex groups having very different body plans) was perplexing. If evolution is true, then the Cambrian Explosion is an implicit admission that the Precambrian rocks don't provide the necessary evidence of the evolution of the major phyla of animals (Figure 8). But since secular science assumes evolution to be true, much collective mental effort is expended to explain the glaring absence of ancestors in the Precambrian. Rather than question the veracity of Darwin's theory, it must be propped up, like the Philistine idol Dagon in its temple. Unfortunately, evidence isn't propping it up, just fine-sounding stories.

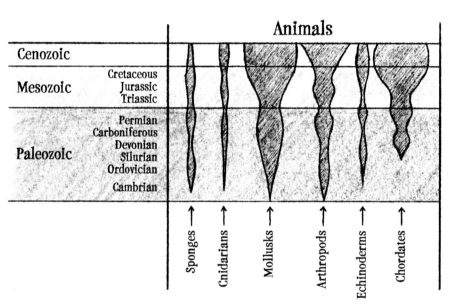

Figure 8. The Cambrian Explosion. The vertical bars represent the fossil record of several animal phyla through geologic time. Time is the vertical axis. If the band extends to the top, the phylum has not gone extinct. Most of them extend back to the Cambrian Period. Before the Cambrian, there were no fossils connecting these phyla to each other or to a common ancestor.

Stephen J. Gould and Niles Eldridge proposed a theory that attempted to account for the absence of these forms. It was called *Punctuated Equilibrium*. This theory essentially maintains that a small isolated population rapidly speciates (here the intermediate forms occur) outside the main population, while the main population is in stasis (not exhibiting any significant changes over time). The smallness of the population is why they are mysteriously absent from the fossil record (they can't be totally absent; just rare enough to not be found in the fossil record.). Gould and Eldridge use the word *equilibrium* to capture the idea of stasis. If this isolated population diverges (speciates) enough to be quite different and the prevailing environmental conditions change such that this population is selected for, then this minority group becomes the winner in the struggle for survival. The main population is selected against and dwindles to extinction. If sedimentation fossilizes the main group *before* the environmental change, and also the new main group *after* the environmental change, the fossil record shows a sudden change in form. Gradualism is obscured and an overview of the fossils shows mostly stasis or *equilibrium*, *punctuated* with seemingly sudden changes in external form. This is similar to telling your teacher that you actually did your homework but it's missing because the dog tore it to bits and ate it. It may be a plausible story but since there is nothing to hand in, you can't prove to your teacher that you did your homework. Punctuated Equilibrium is similar to this in that it may explain the lack of fossils, but it doesn't show that the transition actually happened. It also doesn't explain the genetic basis for any increased complexity. Increased complexity through evolutionary time is just assumed. There is no evidence that any population can increase in complexity from scratch. This has never been observed nor is it theoretically possible (Chapter 6).

In summary, Cambrian Explosion shows *discontinuity* between many phyla in the Cambrian Period and fails to show how all these phyla are connected to a common ancestor in the Precambrian Eon.[21] The rest of this chapter will show that discontinuity is the overwhelming pattern between groups that are family level or higher.

Next I would like to zoom in on the Phylum Arthropoda. Looking at Arthropods in Figure 8, it could be accidentally assumed that this group shows *continuity*. In other words, one could assume there is fossil evidence showing that different arthropods may be connected to each other and to a common ancestor. Not so! Some of you may know that Phylum Arthropoda is by far the largest phylum comprising about 85% of all animals. The vast majority of arthropods are insects (Class Insecta) but other groups include crustaceans, arachnids, millipedes, and centipedes. Class Insecta (about 1 million species) is divided into twenty-nine extant orders. In fact, most of the insect orders appear so suddenly during the Carboniferous Period, it is similarly called the Carboniferous Explosion (Figure 9).

---

21 A thorough and wonderful critique of the Cambrian Explosion is laid out in Stephen C. Meyer's book, *Darwin's Doubt*.

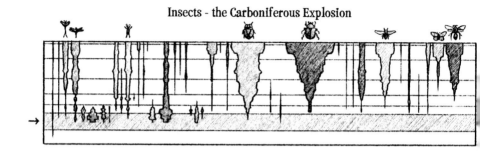

Figure 9: The Carboniferous Explosion. The vertical bars represent the fossil record of several of the insect orders through geologic time. Many of them extend back to the Carboniferous Period. In rocks before the Carboniferous, there are no fossils connecting the insect orders to each other or to a common ancestor.

The largest insect group is the beetle order (Order Coleoptera). The beetle band (about 400,000 species) in Figure 9 is divided into about 180 families. Don't assume there is continuity within the order. There isn't! Created kinds (as discussed in Chapter 10) would be much smaller than order. If we zoom in and examine the beetle fossil record, we can see whether there is continuity between beetle families...or not (Figure 10).

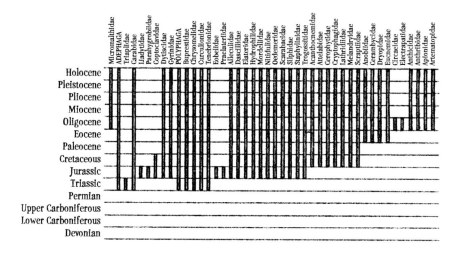

Figure 10: Fossil record of beetles. The vertical bars are just a fraction of the various beetle families. Most extend back to the Mesozoic Era (Cretaceous, Jurassic, and Triassic Periods). But as you can see, prior to these beetle families (listed across the top of the figure) popping up in the fossil record, we don't have transitional fossil beetles connecting these families to each other or to a common beetle ancestor.

I hope you're seeing a trend. We must move on to Phylum Chordata (which includes the more familiar subphylum—the vertebrates). Let me give you some background on this phylum. Chordates all share a stiffening proteinaceous rod, called a notochord, running down the back at some stage in their life. The notochord is replaced by a backbone in the vertebrates. This phylum is divided into three subphyla: the cephalochordates, urochordates, and vertebrates. The cephalochordates (lancelets) are represented by more than 30 species; the urochordates (sea squirts), about 3,000 species; and the vertebrates (the vast majority of this phylum), about 70,000 species. This last group is familiar to everybody including grade schoolers who, early on, learn what a vertebrate is. But here's a quick overview anyway: Vertebrates are the jawless fishes (hagfish and lampreys), cartilaginous fish (sharks, skates, and rays), bony fish (all other fishes), amphibians, reptiles, birds, and mammals. Let's look at the fossil record of these various groups to see if there is continuity or discontinuity (Figure 11).

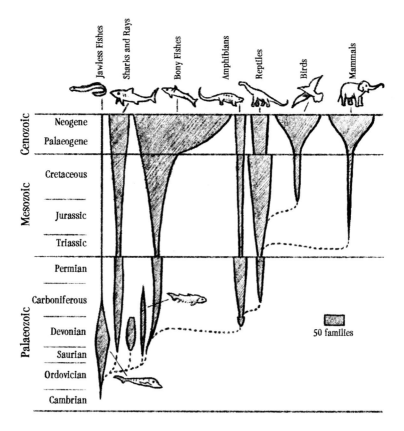

Figure 11: Fossil record of vertebrates. The vertical bars are the classes of vertebrates. The skinny lines or dotted lines represent the presumed intermediate (transitional) forms but they don't actually exist, so we really don't have indisputable fossil vertebrates connecting these classes to each other or to a common ancestor.

What about the disputable intermediate forms? Several famous fossils are hailed as the fish-to-amphibian transition. The group of fish purported to be ancestral to amphibians is the sarcopterygians. The most honored member of this group that is thought to be ancestral to amphibians is *Eusthenopteron* (Figure 12).

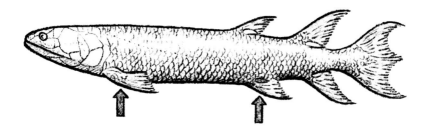

Figure 12: The sarcopterygian fish, *Eusthenopteron*, is considered ancestral to the amphibians because it had fleshy-lobed pectoral fins and pelvic fins. The small bony elements in these fins are thought to be homologous to the limb bones of the first amphibians. Its presence in the late Devonian Period at about 385 million years ago along with the fact that it had lungs made it a good candidate for breathing on land.

The first amphibians of the late Devonian Period (365 million years ago), *Acanthostega* and *Ichthyostega*, are considered the first honest-to-goodness amphibian tetrapods with clearly formed pectoral and pelvic girdles and appendages (Figure 13).

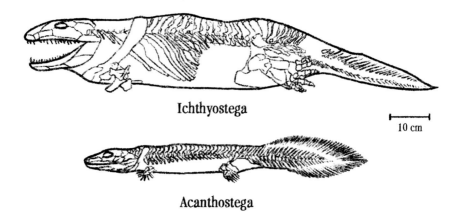

Figure 13: *Acanthostega* and *Ichthyostega*.

The morphological gap between the fish *Eusthenopteron* and the amphibians, *Acanthostega* and *Ichthyostega*, is considerable. In 2004, paleontologists Edward B. Daeschler, Neal H. Shubin, and Farish A. Jenkins Jr., searching purposely in sedimentary rock in northern Canada (Ellesmere Island) that was presumed to be around 375 million years old, finally found a fossil that appeared to be intermediate in form and the right geologic age. It was named *Tiktaalik* (a Inuktitut term for any large shallow-dwelling freshwater fish). It is technically a sarcopterygian fish, but the bones in its fins were more developed than *Eusthenopteron*, and hence it was considered an intermediate form (Figure 14).

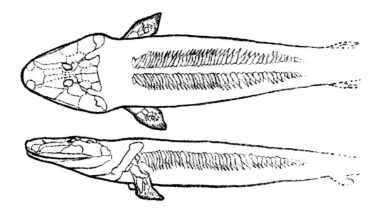

Figure 14: *Tiktaalik*, a sarcopterygian fish having pectoral and pelvic anatomy a bit more developed than *Eusthenopteron*.

These and several other forms are considered transitional and are claimed in technical and popular literature to be indisputable fossil evidence that evolution between fish and tetrapods (four-legged vertebrates) has occurred. What are creationists to make of it? First, we should not try to deny any characteristics that may appear transitional. This is simply disingenuous. Second, we should be much more forthright and describe the fossil for what it is, intermediate features and all. Creationists too often try to shoehorn a fossil find as belonging to this group or that group. In *Tiktaalik's* case, secular scientists have already classified it as a sarcopterygian *fish*. Yes, the evolution community may claim it as an intermediate form, but they can't really prove it. Any such claims are in the realm of historical science. Scientists weren't there nor can they show that this series of fossilized creatures were part of an ancestor-descendant lineage over 20 million years. Creationists believe that God is not beholden to our tidy categories. Most things fit neatly into certain groups, but it is not a problem to creationists when creatures, extinct or extant, "land on the line" between two groups or are a hodge-podge of characteristics like the duck-billed platypus. God made a vast continuum of habitats and microhabitats and He filled them with His creatures (Psalm 104:24). If He created a continuum of habitats, then I would expect God to create a continuum of creatures that fit them. But let's move on.

What about amphibians? Do all amphibian groups show continuity? First, a little background. Class Amphibia includes three extant orders: Order Gymnophiona (the caecilians), Order Anura (the frogs), and Order Caudata (the salamanders and newts). Not only is there discontinuity between these orders, there is discontinuity between the families in each order. Even the supposed common ancestor of amphibians, *Ichthyostega* (at the very bottom) shows a discontinuity between it and other extinct amphibian groups (Figure 15).

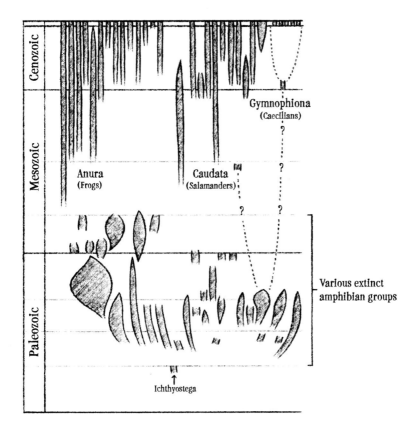

Figure 15: Fossil record of amphibians. The lower groups are a variety of extinct amphibians. The upper half are the modern amphibian groups. No fossil intermediates connect each group to each other or to the common ancestor.

Since the overall pattern is getting redundant I will quickly show you that this pattern exists within all other vertebrate classes. The next figure will show the discontinuity of the reptiles (Figure 16).

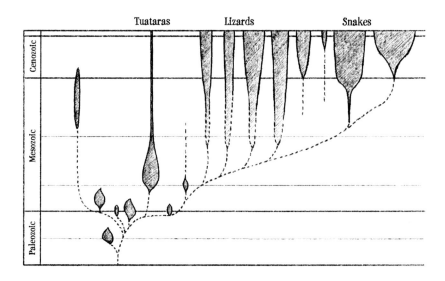

Figure 16: Fossil record of reptiles. Upper center are major lizard groups. The upper right are the snakes. Lower left are mostly extinct reptile groups. Dotted lines represent transitional forms which do not exist.

What about dinosaurs? Ever since the mid-twentieth century dinosaurs loomed large in popular culture, especially with kids. Many go through a dinosaur phase and learn a boatload of different kinds, thanks to dinosaur books and plastic toy dinosaurs. Because of their popularity, this group has served as one of the best ways to brainwash our kids with evolution. If an elementary science book or a children's book on dinosaurs shows an evolutionary tree of dinosaurs, it will look something like the one in Figure 17.

Figure 17: A simplified evolutionary tree of dinosaurs.

But what does the actual fossil data show us? It certainly doesn't look like that. It looks like this (Figure 18):

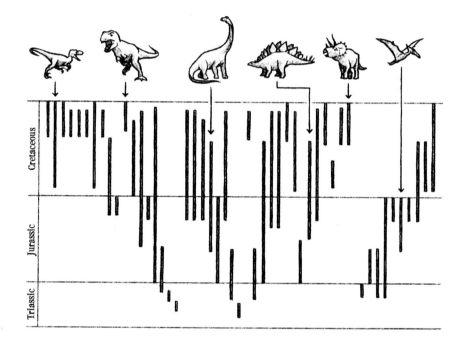

Figure 18: Fossil record of dinosaurs. The vertical lines are the fossil record of dinosaur families. There is no evidence of continuity. There are no fossils connecting dinosaur families to each other or to the common ancestor of all dinosaurs.

Thanks to *Jurassic Park*, even most lay people know that birds supposedly evolved from dinosaurs. The famous *Archaeopteryx* was at first celebrated as a transition (sort of) but it didn't quite fit the perfect intermediate between theropod dinosaurs and birds. Consequently, it has been relegated to an early side branch on the evolutionary tree of birds. Experts do consider it a bird with an assortment of dinosaurian features. The following are its bird characteristics. It had modern asymmetric feathers which indicates that it flew. Its wishbone and small sternum suggest that it was a weak flier. Its feet had curved claws and a hallux (the first digit on the feet of perching birds) which hints that it was a perching, arboreal bird. Its hollow bones hint at probable air sacs and avian lungs. Its movable upper jaw and a large cerebellum were also characteristic of true birds. The following are its dinosaurian features. It lacked a highly keratinized bill and it had teeth in its jaws. Its long bony tail was also more dinosaurian.

Paleontologists consider this more of a mosaic (i.e. a combination of truly bird and truly dinosaurian features) rather than a creature that was intermediate between the two. My view on this is similar to what I said about *Tiktaalik*. Whether it is an intermediate between dinosaurs and birds or a strange mosaic with clear features of both, neither is indisputable proof of evolution. I can see why it pleases evolutionists, but it also pleases God to make strange creatures that don't fit our tidy categories. If you're a creationist disturbed by a bird sporting a number of dinosaurian features, don't be. Don't waste your energy trying to shoehorn this wonderful creature as "just a bird" while underemphasizing its "dinosaurian" features, or "just a dinosaur" while underemphasizing its "bird" features. Embrace it all and praise God for its wonderful weirdness. *Archaeopteryx* having a tail and teeth doesn't make it closer to a dinosaur any more than a Ford having a steering wheel and fenders makes it closer to a Chevy. To an evolutionist, similarity implies common ancestry. To a creationist, it implies common design.

Let's move on to the fossil record of birds where discontinuity is the overwhelming pattern (Figure 19).

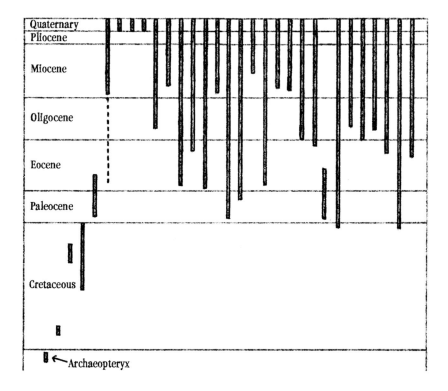

Figure 19: Fossil record of birds. The vertical lines represent the fossil record of orders. If they reach the top they are alive today. Again, there are no fossil intermediates connecting the bird orders to each other or to a common ancestor.

I'm getting bored but I have a few more charts to show you. Bear with me. Figuring out whether an extinct creature is a reptile or mammal is not conclusive because most of the differences are found in the soft anatomy which is usually missing. If a fossil skeleton matches a modern mammal group it's safe to assume it belongs. But if it is clearly a different creature with no living counterparts, we can't assume it was egg-laying or live-bearing or whether it had fur or scales. It becomes pretty speculative so conclusions are highly tentative. With this in mind, let's proceed. Among an extinct, highly disparate group of mammal-like reptiles (Therapsids), it is assumed that the Cynodonts were ancestral to the first mammals. But as you can see, dotted lines prevail (no fossils) between Cynodonts and Mammals and between everything else (Figure 20).

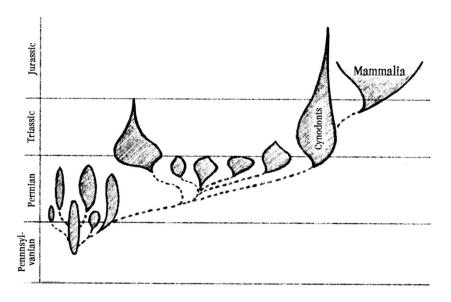

Figure 20: Fossil record of various extinct mammal-like reptiles. Fossil intermediates are missing between mammals and Cynodonts as well as between all the Therapsid groups.

Now we are getting into more familiar territory—the mammals. Without further ado, let us look at the various mammalian orders. This is quite remarkable (Figure 21). Again, there are no intermediate fossils bridging the gaps between these orders or connecting them to a clear common ancestor. In fact, the mammalian "tree" isn't a tree; it's a forest.

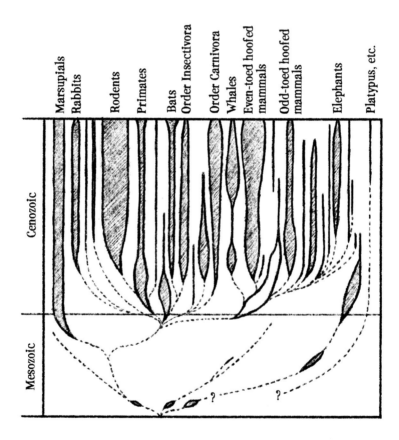

Figure 21: Fossil record of mammalian orders. Only dotted lines (no fossils) inserted due to evolutionary assumptions connect the various orders. Even the presumed trunk doesn't exist.

At this scale, it might be assumed that there is continuity within each order. For example, the vertical bar representing the rodents might imply that all rodents (beavers, squirrels, voles, capybaras, mice, gerbils, etc.) can be traced back to a clear rodent ancestor with fossils connecting them all. For the umpteenth time, not so! Let's zoom in on the rodent fossil record. As you can see yet again, there is discontinuity between all the rodent families, much like the beetle families in Order Coleoptera (Figure 22).

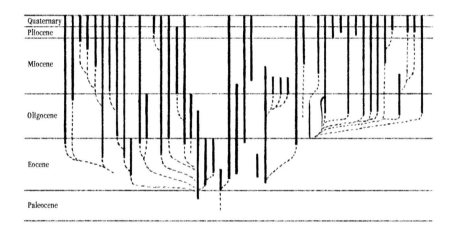

Figure 22: Fossil record of rodent families. The dotted lines inserted due to evolutionary assumptions, makes it into a tree.

To wrap things up, I will switch to another familiar kingdom—the plants. Back in the 1960s E. J. H. Corner, a world-class botanist of the Cambridge Botany School, said something very revealing about the actual pattern of plant fossils. He wrote, "Much evidence can be adduced in favor of the theory of evolution…, but I still think to the unprejuced, the fossil record of plants is in favor of special creation." As you might guess from this quote, the pattern I've been presenting throughout this chapter is found here, too. In Figure 23, we will only look at the discontinuity between major plant groups (mostly phyla). I won't zoom in to show discontinuity between classes, orders, or families (but it is apparent there, too).

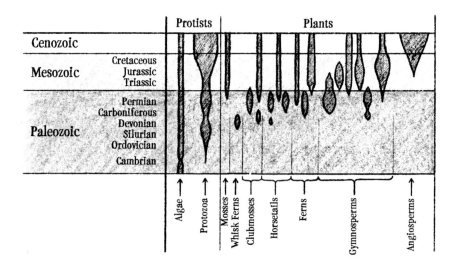

Figure 23: Fossil record of plant phyla. No fossil plants link the plant phyla to each other or to their supposed green algae ancestor.

The take-home message from this chapter is pretty obvious. Regardless of what major group you pick, the fossils are usually easy to pigeonhole into already known groups or subgroups (taxa). If evolution was true, you would expect to find many intermediate forms connecting the subgroups to each other and to the common ancestor of the entire group. In other words, you'd expect to find continuity. This is decidedly not the case. Despite Darwin's hope, discontinuity remains the overwhelming pattern. When strange mosaics like *Tiktaalik* or *Archaeopteryx* are discovered they are the exception, not the rule. The creation model or the evolution model can interpret them according to their paradigm. For example, the creationists can say that God isn't beholden to our tidy categories. He can create all sorts of weird "intermediate" creatures, none of which violates His character or contradicts Scripture. On the other hand, evolutionists can say that evolution produces various dead-end side branches that have a blend of characteristics that hint at what transitional forms may look like. Regardless, these fossil creatures don't prove either model. Scientists modify models to fit (within limits) the actual data. That's just the way it is with historical science.

CHAPTER 6

Who Was Our Ancestor: *Australopithecus* or Adam?

In one sense this could be a continuation of the last chapter, but because we are dealing with the very identity of humanity, I will address it separately. Unfortunately, many people aren't as concerned with whether animals evolved from lower life forms, but the notion that we humans evolved from ape-like creatures evokes (for those who care deeply about their identity and origin) either positive or negative emotions. For those who feel positive emotions, there is happy acceptance. Why? Because they are just products of blind animal evolution. There is no creator to whom they must give account for their behavior. They are simply intelligent animals and therefore they can jolly well follow whatever moral standard (or lack of standard) they like. For those who feel negative emotions, they recoil at the idea of evolving from ape-like creatures. Why? It injures their pride; they are special, created in the image of God, not descended from stupid, dirty apes.

All emotions aside, this is a question about the truth of our actual ancestry. Our autonomy or pride isn't the arbiter of truth. After all, being made from dirt (Genesis 2:7) isn't very flattering either. Humiliating origins does not determine whether something is true or false. If I discover that my great-great-granddad was a horse thief in the wild west, I don't get to deny it because it's an affront to my self-respect. Our ultimate origin is a matter of the authority of God's Word. We should reject human evolution not because it's an insult to our pride, rather because of what Genesis 2:7 says: "Then the Lord God formed the man of dust from the ground and breathed into his nostrils the breath of life, and the man became a living creature." This chapter is to draw conclusions based on the objective assessment of the major hominin fossils without getting into the minutia and history of fossil finds (including the rivalries, the politics, and the selfish ambition) of all the paleoanthropologists of the last century. Simplifying is meant to generate clarity, not to spin the data or sweep inconvenient data under the carpet.

Throughout the twentieth century (particularly the last half of it) there has been a lot of time, money, and effort poured into paleoanthropology (the study of human evolution). Motivations were many and diverse, but the promise of fame to those who found fossil evidence supporting the ape-to-man narrative that Darwin postulated was most certainly a driving factor. Regardless of any sordid or dubious motivations, let's first look at the tidy narrative of human evolution that is spoon-fed (as if it was supported by overwhelming scientific evidence) into students from pre-school to grad school by our public education system. Then we will compare it to the untidy, unclear, controversial fossil evidence. Once this is done, I think you will find that the hominin fossil record is much more in favor of special creation. By their own admission, many paleo-experts[22] think that the human evolution narrative, widely taught in schools, colleges, and universities, is not all that it's cracked up to be. Even though many of their candid comments reveal enormous doubts about the tidy narrative, they still do believe that humans evolved from apes (somehow); they just are acutely aware that the necessary supporting evidence is lacking.

THE SIMPLIFIED STORY

The current narrative (details may vary) in our public school curricula is that the last common ancestor (LCA) of chimps and humans lived between 5 and 10 million years ago (many put it closer to 6 million). There are no fossils of the LCA, but if the theory says we and chimps had a common ancestor, it is assumed to have existed. Around the time of the LCA there was a split that formed two lineages, one leading to chimps and one leading to modern man. The term *hominin* refers to the latter lineage which includes all the supposed intermediate fossil forms between the LCA and man. I won't mention all the fossil hominins that have been discovered to date but will briefly discuss those fossils deemed most important in the line leading to man. The first star of the hominin lineage is *Ardipithecus ramidus* (dated to 4.4 million years ago). The next in line is *Australopithecus afarensis* (Lucy being the most famous specimen assigned to

---

22 Short for Paleoanthropologists—scientists who study fossil hominids.

this species). Lucy is dated to 3.2 million years ago. Another lineage thought to have diverged at about this time led to the robust australopithecines (*Australopithecus robustus* and *Australopithecus boisei*); but since they are considered a side branch not leading to humans, I won't bother you with these brutes anymore. The next transition was the big step up into the genus *Homo* (man). *Homo habilis* holds that place of honor, but as we shall see, this is considered by most paleo-experts as a bogus species (a mixture of *Homo* and *Australopithecus* bones), not a real creature. *Homo habilis* (real or not) supposedly evolved into *Homo erectus* from which arose *Homo neanderthalensis* and finally *Homo sapiens*. In the last twenty years, several newcomers have been presented for inclusion into this family tree. These are *Homo floresiensis* (Hobbit), *Australopithecus sediba*, and *Homo naledi*. Popularizations of this story are often visually portrayed as a linear parade, one form evolving into the next. Evolutionists believe (and we should not misrepresent them) that the human evolutionary tree is a highly branching bush with all but the human branch going extinct.

To lay some philosophical groundwork, we need to realize that we all have certain presuppositions or biases that stem from our worldview. Presuppositions are inescapable; we just want the right ones. So, to start out, it is good to simply acknowledge them and guard ourselves against confirmation bias[23] as we look at the data. Also, as noted before, this is forensic (historical) science. There were

---

23 Confirmation bias—the tendency to search for, interpret, favor, and recall information in a way that confirms or supports one's prior beliefs or values. Definition taken from Wikipedia, https://en.wikipedia.org/wiki/Confirmation_bias, accessed June 28, 2023.

no records from eye-witnesses to testify to the real identity of these creatures when they were alive and walking (or swinging) around. We only have circumstantial evidence and much of it is fragmentary and damaged. Much of the needed skeletal remains are missing so paleo-experts fill in the gaps (more or less) according to their bias. Evolutionary biologists are convinced that humans evolved from ape-like creatures. Creationists are convinced that we were created by God in His image as humans from the start. Even with these differing worldviews, we can still analyze the data without confirmation bias. Many paleo-experts (though firm believers in and committed to the theory of evolution) do just that. They are often brutally honest in drawing conclusions often contrary to what they would like to see.

As I have said before, I don't want to inundate the reader with too much data. Whether we read a detailed assessment or a summary assessment, we won't become paleo-experts. This requires many years of education and training. It often boils down to whom you trust and sometimes that means trusting in people without *confirmation bias* regardless of whether they are evolutionists or creationists.

Ardipithecus ramidus (ARDI)

Discovered in 1992 in Ethiopia, Ardi was given a date of 4.4 million years ago and hailed by its discoverers as the oldest ancestor in the hominin line after the human-chimp split. When announcements like this hit the press, the controversy, doubts, problems, and messiness are mostly overlooked. What people hear and believe is that it is conclusive evidence that human evolution is true. But the public doesn't hear that its remains were highly fragmentary, damaged, unarticulated, and scattered about the countryside. As a result, reconstructions or claims as to its honorable hominin status are decidedly dubious and largely a matter of faith. Objective assessment of the actual evidence of Ardi indicates that the creature was some sort of extinct ape. Why? Its hands and feet (containing opposable halluces or big toes) were ape-like, its arms and legs had proportions typical of an ape, and there was no solid evidence that it walked upright. Those advocating (Ardi's discoverers) its hominin status received much push-back from the paleo-community. The paleo-experts who doubted Ardi's hominin status claimed that the traits used to place it in the human line are features possessed by apes alive today. Whatever Ardi was is highly speculative and now should be a moot point. The problematic find in 2017 of human footprints on the island of Crete with a date of 5.6 to 5.7 million years ago makes Ardi irrelevant by their own reckoning.[24] If modern humans were walking around Crete over a million years before Ardi lived and died in Ethiopia, then Ardi can't be the beginning of the human line. In fact, none of the other hominin species can be ancestral to humans either. You either have to conclude that the footprints aren't human, or the date is way off, or you throw out the entire precious hominin evolutionary tree.

---

24 Gerard D. Gierliński et. al., "Possible hominin footprints from the late Miocene (c. 5.7 Ma) of Crete?", *Proceedings of the Geologists' Association* 128, no. 5–6 (October 2017): 697–710, https://www.sciencedirect.com/science/article/pii/S001678781730113X

Australopithecus afarensis (Southern Ape of Afar— Lucy and Kin)

There is much interesting history surrounding *Australopithecus afarensis*, but here I will try to cut to the chase. Lucy happens to be an individual skeleton that was discovered by Donald Johanson at the Hadar site in 1974 in the Afar region of Ethiopia, hence its specific epithet *afarensis*. Lucy was the best (though incomplete) representative of the species. The rest of the species is represented by an assortment of bones and bone fragments (assumed to be Lucy's kind). These have been helpful in ascertaining the nature of some of Lucy's missing parts including the hands and feet. Nevertheless, what does the fossil evidence of Lucy's kind tell us? Lucy was assigned a date of 3.2 million years ago, although *A. afarensis* (Lucy's kind) is thought to span from roughly 4 to 3 million years ago. After thorough examination of all the skeletal evidence of *A. afarensis*, many paleo-experts (including Richard Leakey) admit that it appears to have all the features of an ape. They have honestly acknowledged that *A. afarensis* is very ape-like in the following features: skull, jaw, and face size, cranial capacity, rib cage, hands and feet, spine, hips and shoulder, limb proportions, and knee joint structure. There are indeed some of Lucy's bones that are larger and more human-like…and they might well be. *Homo* bones (indistinguishable from *Homo sapiens*) have been found in the same bone beds at the Hadar site where Lucy was found. Again, the footprints (indistinguishable from human footprints) found on the island of Crete are over 2 million years older than Lucy. The Laetoli (Tanzania) footprints (also indistinguishable from human footprints) are older by about a half a million years. If we take these footprints at face value (if it had human feet and walked like a human, it probably was a human), then *A. afarensis* can't be the ancestor of humans if they appear in the record much more recently than humans do. For a more detailed assessment of the actual evidence, I highly recommend *Contested Bones* by C. Rupe and J. C. Sanford.

Homo habilis (Handy-Man)

All the king's horses and all the king's men put *Humpty habilis* together...wrong. To make a long story short, the bones and teeth that came to be known as *Homo habilis* were discovered by Mary Leakey and her son, Jonathan, at Olduvai Gorge in Tanzania in 1960. Sixty years of paleo-experts researching this find has only served to make its identity more murky and doubtful. An objective assessment of *Homo habilis* is that it appears to be the product of piecing together a mixed assortment of a few scattered and damaged bones and fragments belonging to two different creatures, the small creature being the more ape-like (*Australopithecus*) and the larger, more human-like (*Homo*). To be fair, it is exceedingly difficult (if not impossible) to accurately construct two different skeletons with each piece placed in the correct skeleton when you're picking them from a scattered assortment in a mixed bone bed[25] with bones missing. As a thought experiment, consider accurately reassembling broken and torn puzzle pieces from several different jigsaw

---

25 Having bones from several different species.

puzzles dumped and mixed up together (over a large area) with many pieces missing from each puzzle. Each bone or bone fragment is a puzzle piece. Each piece isn't labeled *Australopithecus* or *Homo* so we don't know for sure which creature they belong to. But it gets worse. Add to this mess the desire to find an intermediate form bridging the gap between *Australopithecus* and *Homo* and you get an idea of how easy it would be to assemble a creature with bones from both. Consequently, any conclusions about its identity should be viewed through the squinty eyes of dubiousness. Often the researchers say as much. With this high level of uncertainty, many honest paleo-experts have agreed that *Homo habilis* is a man-made, made-up ape-man. Paleo-experts (for whatever reason) cobbled together ape and human bones (acknowledging that it may be just that) and allowed the press to hail it as a missing link. Even though the experts knew their claims were uncertain, they didn't seem to mind when the press, textbook writers, teachers, and professors dogmatically presented it to the public and students (grade school though grad school) as undeniable evidence for human evolution. This is simply not OK.

Homo erectus (Erect Man)

As with all these other finds, *Homo erectus* has a long and interesting history; but if we ignore the narrative that evolutionists want it to conform to, its morphological characteristics fall within the range of *Homo sapiens* (modern man) and therefore it should be considered as such. Around 300 specimens have been found on several continents that have been assigned to this species. Because its average cranial capacity is smaller than the average for *Homo sapiens*, it is considered to be a less advanced hominin species and is held to be the ancestor of *Homo sapiens*. It also had heavy brow ridges which gave it a more brutish aspect. However, its postcranial (non-skull) skeleton is indistinguishable from *Homo sapiens*. Paleo-experts that are considered "lumpers"[26] conclude that *Homo erectus* should really be lumped with *Homo sapiens*. Based on the range of differences that we see in all modern *Homo sapiens*, the lumpers are making the correct call. There is no evidence that warrants making them a different species. Though the average cranial capacity is smaller than *Homo sapiens*, they fall within the range of *Homo sapiens*' cranial capacity. Cultural evidence surrounding *Homo erectus* finds indicates that they were intelligent, reasoning creatures capable of language, boat-building, sailing, jewelry-making, knot-tying, fire-making, cooking, art, woodworking, hunting, caring for the infirm, and other skills associated with humans. The small body size (the Hobbit) and smaller than average skull size are easily explained by several factors often associated with small isolated populations. One factor is malnutrition. Reduced body and especially brain-size are adaptive to hunter-gatherers who frequently failed to put food on the table (or rather cave floor) and faced starvation conditions. Those with smaller bodies and brains were naturally selected for. The food they did get would more likely supply their reduced caloric requirements. They were less likely to starve

---

26 Lumpers consider minor skeletal differences insufficient to classify a find as a distinct species, while splitters tend to consider those same differences sufficient to classify a find as being a distinct species.

to death than their bigger hungrier brethren. This is called *reductive evolution* (evolution that subtracts rather than adds). Another factor that small isolated populations faced was *inbreeding depression*.[27] All these factors are acknowledged by paleo-experts, so the idea that these creatures were evolving into man (almost, but not quite) is not the best or most reasonable explanation. Lastly, the dates for all the *Homo erectus* material ranges from 1.9 million to 140,000 years ago. As is the case for all hominins dated younger than 5.7 million years old, whatever they may be, they can't be ancestral to *Homo sapiens*. If *Homo sapiens* footprints were squished into the mud on the island of Crete 3.8 million years before the earliest *Homo erectus*, then it's a moot point.

Homo floresiensis (The Hobbit)

In 2003 the skull and partial skeleton of a diminutive hominin were discovered on the island of Flores, Indonesia and given the nickname "the Hobbit." There is much interesting history here too, but since this chapter is already too long, I will bottom-line it. Given their similarities to pygmies of the area today, coupled with evidence that shows their cultural attainments were on par with *Homo sapiens*, it is most reasonable to place them there. The majority of paleo-experts place them in either *Homo sapiens* or *Homo erectus*. Since the latter should be subsumed under *H. sapiens*, the case for the hobbit being subhuman is weak. As already described under *Homo erectus*, the small cranial capacity and

---

27 Physical or mental defects or pathologies resulting from inbreeding.

other abnormal features may be due to *reductive evolution* and *inbreeding depression* rather than being some pre-human evolutionary intermediate.

Australopithecus sediba

Another contender for the honored transition between *Australopithecus* and *Homo* was discovered in 2008 in South Africa and dubbed *Australopithecus sediba*. After much study it appears that an actual "intermediate" doesn't exist. The bone bed where it was discovered was a jumbled mix of ape, human, and other animal bones. As already mentioned, when zealously seeking to find the missing link it is understandable to accidentally piece together bones and bone fragments from two different species to create something of an intermediate. This is essentially the same mistake that led to *Homo habilis*. According to brutally honest paleo-experts, who are still committed to evolutionary theory but refuse to make claims contrary to an honest assessment of the data, *Australopithecus sediba* is really a composite of ape and human bones.

- Human → hand bones, hip bones
- Ape → skull/brain case, rib cage, upper limb bones (humerus, ulna, radius, clavicle, and scapula)
- A mix of man and ape → spine and jaw

Australopithecus sediba is a product of wishful puzzle building (not necessarily with deceitful intent), of putting together a puzzle by accidentally drawing pieces from two different puzzles. Putting together a puzzle wrongly may be forgivable under some circumstances; but when the researchers know that the bone bed is a mixed mess and they know there is glory to be gained if a transitional hominid is discovered and they have a strong evolutionary bias, the idea of an innocent mistake rings hollow.

Homo naledi

A bone bed of many individuals of the same species was discovered in 2013 in a cave system in South Africa. It was named *Homo naledi* and was acclaimed as a new hominin species that was almost human. Although classified in the genus *Homo* (because most of its skeleton is clearly human) its discoverer Lee Berger found that it had a few *Australopithecus*-like features including its small cranial capacity, flared rib cage, scapula orientation, curved finger bones (which suggests a tree-climbing existence), etc. It is well known that certain living conditions and behavior can mold portions of the skeleton in different ways that resemble some traits found in *H. naledi*. Also, as mentioned before, isolated populations result in inbreeding depression and malnutrition. These both can lead to reductive evolution resulting in reduced cranial capacity and other skeletal pathologies. Because of these very likely living conditions, the evidence that *H. naledi* is another almost-human hominin is sketchy at best.

Of course there will continue to be more fossil discoveries, but my prediction is that we will hear more of the same. Based on all the fossil finds to date, the evidence is far from convincing. Once we hear the paleo-experts' uncertainties, doubts, and qualifications, the narrative of human evolution starts to sound more and more like a fabricated story built from an ideological (and emotional) commitment to evolution, rather than from actual evidence.

CHAPTER 7
Alphabet Soup—Can Biological Building Blocks Become a Living Cell by Chance?

I have lectured on this topic for many years and it never ceases to amaze me that intelligent scientists actually believe the myth of the primordial soup. I can easily understand how the average Joe, who buys the *National Enquirer*, can be duped into believing this myth; but when the former group believes it, that kind of credulousness seems to be as inexplicable as a Ph.D. in developmental biology writing a grant proposal to investigate the tabloid headline "Teen gives birth to Bigfoot Baby." The difference between the proposition that bacteria-like cells formed by chance in an early Earth soup and a teen giving birth to a Bigfoot baby is that the former is much more ridiculous. Why then do they believe it? There are probably many reasons why educated people actually believe this, but here are a few for starters: 1) The proposition is found in textbooks couched in technical, scientific language written by authoritative academics; 2) the reader trusts these authors because they are presumably experts and have Ph.D.s; 3) the reader does not exercise critical thinking when reading it; and

4) they are inclined to believe any ideas that explain away God, which is always handy in ridding their minds of moral obligations and the guilt that accompanies disobeying those obligations. If the Bigfoot baby was explained in plausible, technical terminology in a textbook, I bet you would get a lot more believers. College students are often just as gullible, but to successfully dupe them, the deceiver must be subtle, clever, educated, self-deceived (or just plain evil), and lastly, should promulgate the deception using a different medium—popular science websites, technical journals, and textbooks.

WHY IS THIS TOPIC IMPORTANT IN APOLOGETICS?

Since the publication of *The Origin of Species*, the evolutionary tree has been thoroughly inculcated into all facets of society, from grade school to grad school, from the soft to the hard sciences, from signage in national parks to entertainment in *Jurassic Park*. It has been so established that the evolutionary tree of common descent is household knowledge. If I want to cut down this evolutionary tree (which I do) then the most sensible place to start chopping is at its very base. The base of this tree is the chemical origin of life, also called *abiogenesis*, occurring in the primordial soup. As we shall see in this chapter, the primordial soup in which abiogenesis supposedly occurred has no geological evidence that it ever existed nor can it theoretically generate the necessary macromolecules if we assume it did exist. Even if we granted them a generous handicap—a soup composed of useful macromolecules—it could still never auto-assemble itself into bacteria-like cells. This myth can be theoretically and empirically shot down at every step along the way. But before I dive into the primordial soup topic, I want to give a warning to certain people who are drawn to books like this. Many people who read evolution-creation books often want a stack of evolution-refuting trivia that they can memorize and then hose down some hapless evolutionist with it at the slightest provocation. They don't really want to understand the material; they just want to pummel an evolutionist with it. This lack of understanding not only discredits creationism and the Gospel with misrepresentation and misapplication of truth but it also results in the reader rejecting true facets of evolution simply because evolutionists believe it. Consequently, I want to carefully educate with principles that will truly help you 1) strengthen your own faith, 2) understand the fundamental errors in evolutionary thinking, 3) be able to intelligently show an evolutionist the logical and scientific fallacies implicit in the theory, and 4) make subtle true/false distinctions within the General Theory of Evolution. Since evolution is predominantly a biological theory, it requires that you know some things about basic biology that at first do not seem relevant to the argument.

Importance and Complexity of Proteins

The cell is an exceedingly complex array of molecular machines and hundreds of chemical reactions working together as an integrated whole. It's like a miniature factory precisely regulating all imports of raw materials and exports of its products and wastes. It also contains all the instructions in the DNA for the construction and job descriptions of its factory workers and machines (the factory workers are proteins called enzymes and most of the machines are also proteins).[28] This means that the factory workers aren't hired on from the outside; they are made precisely when they are needed within the cell and their specific job is not determined by training but by the protein's three-dimensional shape. You may ask, what determines the 3D shape of a protein? To answer that question, I need to give you a smidge of biochemistry.

Proteins are large molecules made of smaller molecules called amino acids. There are about twenty different kinds of amino acids used in proteins. Amino acids are hooked together to form proteins like pearls are strung together to form a necklace. We are familiar with tools like hammers, saws, crowbars, and screwdrivers. These tools do certain jobs because they are crafted into certain shapes by certain casts that were designed by certain manufacturers. Proteins (including enzymes) follow the same pattern. Each protein does a certain job because it has a certain shape. That 3D shape is constrained by the sequence of amino acids. This is analogous to the importance of the sequence of digits in a telephone number. If you punch in the wrong number, you either get nothing or a different person on the other end. In the same way, if the sequence of amino acids is wrong, the protein may fold up differently resulting in a slightly to radically different 3D shape depending on

---

28 Some biological codes needed for life lie elsewhere; not just in the genome.

how many amino acids are different or out of place. The proper functioning of a cell requires hundreds of proteins performing hundreds of different processes and reactions. In order for everything to operate properly, every enzyme, structural protein, or machine-like protein requires that they all are built with the correct sequence of amino acids, and are made at the right place, at the right time, and in the right quantities. To take it a step further, the sequence of amino acids in each protein is determined by the sequence of nucleotides within DNA. DNA, as many of us know, is the genetic blueprint of life, but more specifically, it contains the genetic blueprint for each protein. Gene is a word that we are all familiar with but only specialists seem to know the definition. A gene is a segment of DNA that codes for one or more proteins. In order for a cell to manufacture all the proteins necessary for cell function, each gene (blueprint) coding for each protein must somehow be "read" (like any other code) and translated into specific proteins. These proteins are not individualistic. They work in concert with each other and often the function of one protein is dependent on the presence of many other properly manufactured proteins.

I'm just scratching the surface of cell function. It is vastly more complex and the information content is staggering. I'm describing the cell very simplistically using very broad brush strokes. Nevertheless, what I have expressed is accurate. In order for the simplest cell to form by chance, hundreds of proteins need to form by chance (not to mention a whole array of other molecules, i.e., phospholipids, DNA, carbohydrates, coenzymes, etc.) at the same time, in the same place, and bound by a membrane. In order for this primitive cell to have even a slight chance of surviving, each protein would have to auto-assemble itself without the help of the highly sophisticated and efficient ribosomal assembly-line machinery present in modern-day cells. Each protein would also require a highly specific amino acid sequence necessary for its specific 3D shape and its specific function. It's not just enough for one functional protein to form by chance. As I said before, cell processes are highly integrated and work together like an athletic team or parts of a sophisticated machine, thus proper function is inextricably linked to the proper formation of other proteins or biological molecules. If a quarterback formed by chance (doubtful, but work with me here), that doesn't produce a football game. For example, helicase is an enzyme that unzips the double helix of DNA to facilitate DNA replication. If helicase happened to form by chance in the primordial soup, it would be totally useless outside the context of DNA replication and all the proteins and enzymes that play an essential part in this amazing molecular dance. Most cellular processes are highly coordinated and regulated systems of structural and functional proteins. I will address this further in the following chapter, summarizing Michael Behe's excellent thesis on irreducible complexities, but for now I would like to belabor the point that even one protein (whether it could be a functional component in an integrated cellular system or not) is simply an event that *would not happen* apart from a cell with the necessary information and machinery to construct it. Obviously mechanistic evolutionists think it *did happen* through naturalistic processes, so I will first summarize their hypothesis and then explain why it is both untenable and ludicrous at every step.

The Oparin-Haldane Hypothesis

The specific name for this well-known version of chemical evolution came from two men who outlined a "plausible" scenario of how the first cells could have arisen from inorganic chemicals on the surface of the early earth 3.8 billion years ago. A Russian biochemist, Alexander Ivanovich Oparin (1924) and a British biologist, J. B. S. Haldane (1928), outlined the basic hypothesis but it was later elaborated on by a couple of other mechanist thinkers (J. D. Bernal and Harold Urey in the mid-1900s) who contributed possible mechanisms for this hypothesis. The basic outline is quite simple which I will break into six big ifs:

1. The early earth atmosphere was reducing which means virtually no oxygen (i.e., organic compound friendly).

2. Inorganic compounds assumed to be present in the atmosphere were CH_4, H_2O, H_2, and NH_3.

3. Energy sources such as UV light, geothermal energy, lightning, shock waves, etc., caused reactions to take place between the above chemicals that produced a variety of organic compounds. Among these compounds would be the important building blocks for the major biological molecules (lipids, proteins, carbohydrates, and nucleic acids).

    a. fatty acids and glycerol → lipids

    b. amino acids → proteins

c. monosaccharides ⟶ polysaccharides

d. nucleotides (each nucleotide is composed of a nitrogenous base, sugar,[29] and a phosphate) ⟶ DNA or RNA

4. These various building blocks polymerized (linked together) into the important biological molecules shown to the right of each arrow in step 3.a–d.

5. While these were forming, they were happily enclosed within biological membranes (from 3a) to form protocells. These protocells would then await further random events that would increase their complexity.

6. Eventually these molecules would somehow become pre-adapted to each other such that they would begin to perform rudimentary metabolism. DNA, which formed by random linkage of nucleotides, somehow found itself with nucleotides ordered in such a way that, if given the proper structural machinery and enzymes (which all happily formed by chance), could dictate the order of amino acids of the newly made proteins. Of course these proteins would be better equipped to do metabolism because they had a specified sequence of amino acids determined by the random sequence of DNA. Make sense?

---

29 A five carbon monosaccharide: ribose or deoxyribose.

The Stanley Miller Experiment

It wasn't until the 1950s that the Oparin-Haldane hypothesis was tested. One of Harold Urey's grad students, Stanley Miller, set up an apparatus (Figure 24) that was to simulate the early earth atmosphere.

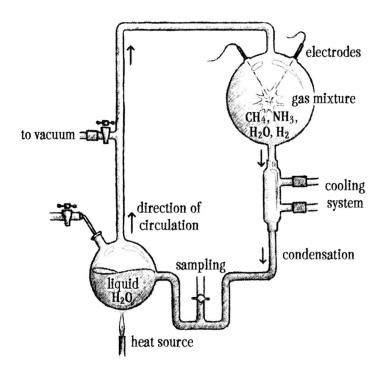

Figure 24: Miller-Urey apparatus.

The boiling helped to circulate the various reactants which were added. These reactants were hydrogen gas (H_2), methane (CH_4), water (H_2O), and ammonia (NH_3).

Of course, they needed some sort of energy source to catalyze the reaction, so a spark discharge chamber (with electrodes) was included in the apparatus (the sparks represented early earth lightning). When the gases (reactants) entered the spark discharge chamber, the spark catalyzed reactions. The products then descended into the condenser and then into the trap at the bottom of the apparatus (this represented the descent of the products generated by the lightning into the primordial ocean). The trap could then be sampled and analyzed to see what was generated. Lo and behold, five different kinds of amino acids found in proteins were produced along with a variety of other organic compounds. The media, true to form, heralded these results to the public with such a spin as to convey that Miller was close to demonstrating that chemicals and energy could form life in the lab. Much rejoicing ensued among those scientists eager to show how life could form naturalistically. Other researchers, spurred on by Miller's success, jumped on the bandwagon to try variations on his experimental design. These researchers set up similar apparatuses but they varied the ingredients Miller used (to a greater or lesser extent) as well as the energy sources (instead of electricity they used heat, UV radiation, shock waves, or high-energy chemicals). The hope was to see if the list of products could be lengthened. In short, they didn't get anything more than a random array of organic chemicals. Yes, some researchers had a longer list of amino acids and other organic chemicals than Miller but eventually the excitement of these experiments began to wane. It became all too clear that naturalistic processes (laws of chemistry) were not going to link these amino acid building blocks together in any biologically meaningful way. In other words, they weren't forming proteins.

Roadblocks Preventing Protein Formation

If you have any chemistry background, especially organic chemistry, you are probably aware that energy sources that generated these building blocks are proverbial swords that cut both ways. Ultraviolet light, heat, shock waves, and electricity do indeed generate amino acids, etc., but in unregulated and uncontrolled quantities, they are just as effective in destroying the products they are capable of forming. Actually, if taken all together, these energy sources will have a greater destructive rather than constructive effect. This destructive synergism is called the *Concerto Effect*.

Given the best case scenario, a tidal pool loaded with amino acids and a host of other organic chemicals cloistered away from destructive radiation in a coastal cavern *will never make a protein*, let alone many proteins. Proteins are built according to genetic blueprints. But this is not all. There are also a lot of cellular factory-workers (enzymes) and fine-tuned cellular machinery (ribosomes, endoplasmic reticulum, etc.) needed to successfully manufacture these amazing molecules. What the materialists must believe is that unguided laws of chemistry can accomplish what only a highly-informational complex cellular factory can. Amino acids, like all carbon compounds, have *functional groups* which have the capacity to react with other *functional groups* on other compounds, not just other amino acids. Without the assembly-line infrastructure of the cell, amino acid *functional groups* would react indiscriminately with any appropriate *functional group* regardless of whether it was on an amino acid or not. In other words, countless directionless *side reactions* could occur that would not lead to a properly constructed protein. For instance, the letter-shaped noodles in Campbell's Alphabet Soup do not spontaneously form a Shakespearian sonnet from the random thermal currents, buoyancy, and adhesive properties of cooked noodles. Noodle letters adhere to other ingredients in the soup such as floating vegetables. They don't just adhere to other noodle letters. If you came to the table and found a Shakespearian sonnet floating on the surface of your soup, you would rightly conclude that an external intelligence (you may dispute its level of intelligence but my point is any external intelligence not contained in the soup) took considerable effort to arrange the letters in an intelligible pattern for your amusement. Why? Although you know that noodle letters stuck together to form a sonnet on the soup's surface in no way defies the properties of noodle and soup chemistry, you also intuitively know that this kind of informational arrangement has such an infinitesimally small probability, you simply conclude your prankster roommate has too much free time on his hands. For you to conclude that this event happened by chance simply because the noodles were floating and adhering in ways that were not miraculous is simply irrational. You say, "There's nothing miraculous; they're noodles floating in the soup." True, but it's not *just* noodles in the soup! Although there is no additional material present in the soup, information from an intelligent source has been superimposed on the soup using noodles. Any human that can read intuitively knows that.

To continue with the Concerto effect, it is known that given the environmental conditions present on this hypothetical earth, most of the amino acids that formed in the upper atmosphere would be annihilated while running a gauntlet of destructive energy sources during their descent to the ocean. Those amino acids that did make it would be few and far between, and they would be so diluted that linking reactions would not happen at adequate levels. Getting back to the soup analogy, it would be like a few noodle letters so widely separated on the surface of a large lake of soup that they rarely, if ever, come in contact. Chemically speaking, lower concentrations also tend to encourage molecules to disconnect rather than connect with each other.

For argument's sake, even if we assume that amino acids were highly concentrated and were somehow prevented from reacting with other chemicals, the chemical roadblocks are still impassable. First, amino acids can react with *functional groups* that would generate wrong linkages between amino acids. In the soup analogy, it would be like noodle letters adhering to each other not side against side (like normal writing) but side against top like this (Figure 25).

Figure 25: Alphabet soup.

Again, given the best handicap (boosting it over every aforementioned roadblock), if we grant a tidal pool loaded with amino acids, cloistered in a coastal cavern away from nasty destructive radiation, never reacting with other chemicals, only reacting with other amino acids at the proper *functional groups* to form peptide bonds, it will still never make a functional protein. Why? The amino acids, even if they did link spontaneously making peptide bonds, would form a *random sequence* of amino acids leading to a functionless protein.

Whether it is binary code in computer language or letters in a human language, information is dependent on a specified sequence of characters. You can't just randomly type in 1s and 0s and hope to generate useful software nor can you throw pea gravel at your keyboard continually for 24 hours and generate a meaningful book. Likewise, concentrated noodle-letters floating in soup, even if it is devoid of any other floating food, cannot generate Shakespearian sonnets. For fun, let's just assume the impossible happened: a protein with a known biological function was formed by the random collisions of the amino acids in the primordial alphabet soup. What is it going to do? In order for a protein to do anything useful *in a cell* it must be *part of one*. This means that it has to perform its function in the context of a larger integrated system. This usually involves hundreds of other proteins and other molecules. If it forms all by itself, it will be as useless as a carburetor forming by chance in the heart of a geologically active mountain. As impossible as that is, it is still a dead end street. The carburetor is meaningless and functionless unless it is integrated in the context of the appropriate engine.

Time doesn't help either. By the time the pistons and cylinders form by chance (several million years lapse between each auto part forming) in the same mountain, the carburetor has disintegrated. Of course a lot more than those three car parts need to form by chance at the right time(s) in order for a brand new car to roll out of a steaming fissure in the side of this mountain. For a cell to even live and survive the duration of one day, hundreds to thousands of proteins (plus DNA, RNA, etc.) must form by chance at the same place, at the same time, and in the right quantities, enclosed in a functional cell membrane with all its proteins integrated so that it is metabolically alive, able to obtain energy and raw materials from its surroundings, able to rid itself of waste, and able to divide. Let the primordial soup stew and brew for a billion years given a best-case scenario and still the simplest protein won't happen…let alone a cell. How do I know without setting up an experiment to test this bold claim? Certain claims don't need further testing. In every human-generated software system, we all know that it was generated by intelligence. It would be a huge waste of time and money to set up an experiment to calculate the probability of its chance formation. Information contained in DNA encoding the sequence of proteins in the simplest of cells is a very information-rich system. Why do our mainstream scientific authorities insist that it was generated by chance? It is not because the evidence points to the sufficiency of chance, time, and energy. Rather, the evidence is overwhelming that all biological life was designed, so why do they insist otherwise? Answer: The Designer/Creator has remained out of reach of our empirical tools and therefore since science cannot directly observe or test any supernatural Cause that transcends nature, they conclude that the ultimate cause must be nature. The hubris of the scientific community hates not being able to observe or test ultimate causes using man-made methodologies. Therefore, it is easier for them to simply deny any cause that transcends nature. Moreover, if a transcendent Creator does exist, all sorts of troublesome implications and questions crop up. To list a few: Why am I here? What if the Creator has expectations about how I behave? Why do bad things happen if the Creator is presumably good? But these are religious questions. You shouldn't reject the best explanation simply because you might be uncomfortable with certain religious implications when the intuitive answer to all this complexity is obvious.

CHAPTER 8

Micro-machines—Is a Darwinian Origin of Irreducible Complexities Possible?

When King David wrote in the oft-quoted 139th Psalm, "I am fearfully and wonderfully made, I know that full well," it was clear to him that the human body wasn't a product of chance. Paul echoes these thoughts in Romans 1:20: "For since the creation of the world God's invisible qualities—his eternal power and divine nature—have been clearly seen, being understood from what has been made, so that men are without excuse." No one needs to be a cell or molecular biologist to ascertain design in the human body. You don't need to be a car mechanic to know that automobiles are designed by automotive engineers. That is intuitively obvious. How much more a living cell? The more you know the machinery of biological systems at the tissue and cellular levels, the more you can powerfully demonstrate the absurdity of any unicellular critter (or even a part of a unicellular critter) arising from the alphabet soup regardless of how long a time frame is granted.

Prior to the discovery of DNA's structure by James Watson and Francis Crick in 1953, cell biology was largely descriptive. After their discovery, cell biologists started in earnest to ask "why" questions and began developing and fine-tuning the procedures to answer them. Over the last several decades countless biologists have teased apart countless cell types and have unveiled many of their inner workings. We are now well into this molecular revolution and now know that the phrase "exceedingly complex" to describe a living cell is an egregious understatement. This unveiling of the cell has demolished the loosely held idea that cells were simple blobs of protoplasm. Instead, we find that cells are loaded with genetic information (DNA) contained within the nucleus. Think of this DNA as a cellular CD with hundreds to thousands of files. There are teams of proteins that work together to open those files (genes) and print out hard copies of each document when needed. Each printed document is a protein. In this loose analogy, the printer is the ribosome. The analogy breaks down in the fact that the protein is not just a hard copy of the information. Rather, it folds up into a functional entity (like interactive origami) that then goes off to fulfill its role in the life of the cell. Each cell also must be able to regulate what files remain closed and what files get opened and printed. You don't want your liver cells to open the file (the gene) that codes for hair protein. I don't think you'd be keen on a hairy liver. The cell also has an internal framework of proteins to give it support and shape. These proteins would be analogous to tent poles in a modern dome tent. Some of these proteins not only provide support but also form "gondola cables" for cellular transport. Little molecular motors (also proteins) attach to these cables and ratchet along them to "ferry" cellular supplies to and from different parts of the cell. The cell is fluid-filled so all these processes and other things I have mentioned are submerged in a densely populated aquatic environment (cytoplasm). Since the cell can be compared to a minute factory, it requires raw materials to be brought inside the cell in the right quantities. It also requires a way of excreting waste and exporting useful products. These little factories must also have an energy supply—their own power generator. Molecular batteries called ATP are produced inside the cell's tiny power plants called mitochondria. Usually there are dozens of mitochondria that burn fuel brought into the cell. This fuel is what we call food. These mitochondria burn this fuel (using enzymes) which converts food energy into a form usable by the cell (ATP). When we burn any fuel such as oil, gas, wood, or coal it produces energy. If we harness the energy, we can use it to do mechanical, electrical, or chemical work. If it isn't harnessed, most of the energy is lost as heat or light. Our mitochondria harness the energy released in this "burn" to generate these ATP batteries. These portable batteries then expend their stored energy on various cell activities that require it. When they expend their energy, they need recharging so they cycle back to the mitochondria to get re-energized. Hence the need to keep eating. Of course a living cell wouldn't get very far if it couldn't divide. Normal cell division is called mitosis and cytokinesis. Although many of us vaguely remember this word as representing a dry factoid from some ancient biology class, it is nevertheless amazingly complex. Cell biologists are still trying to figure out the details. It requires hundreds of different proteins which carry out their duties with mindless precision. I say mindless because, if you recall from Chapter 7, a protein's function is determined by its three-dimensional shape. A protein's 3D shape is determined by how it folds up. Paper can be folded according to an artistic pattern (origami) or can be randomly wadded up (paper trash). The 3D shape of a protein is not only constrained by the ordered sequence of amino acids but also by chaperone proteins (which help fold it into its proper shape). The Mind that designed these mindless proteins is the mind of God Who designed the DNA that codes for the precise sequence of amino acids as well as the chaperone proteins (origami experts).

I have done a little teaching but my goal here is not to give you an introductory biology course. Rather, I want to give you a feel for the dynamics of cell life and then summarize a water-tight argument for design that was presented in the book *Darwin's Black Box*. This book, which I highly recommend, was written by a biochemist and friend, Michael Behe, and over the past decade has caused a number of staunch Darwinists to pitch hissy-fits over his conclusion. The main thesis, which he develops thoroughly, he calls *irreducible complexity*. To quote Behe's definition, an irreducible complexity is "a single system composed of several well-matched, interacting parts that contribute to the basic function, wherein the removal of any one of the parts causes the system to effectively cease functioning."[30] This concept was so well-developed and articulated that it really threw a monkey wrench in the Theory of Evolution.

---

30 Michael J. Behe, *Darwin's Black Box* (New York: Free Press, 1996), 39.

The problem is simply this: in order for some structure or function to be added to a creature's body, it needs to confer some advantage that enhances the recipient's chance of survival. If you look at the definition closely, you will notice that it can't be added to a creature's accouterments via the "installment plan." Each part is useless unless it is in the context of the complete functioning system (see Chapter 7). In other words, the whole system must be added to the evolving creature in its entirety or it won't work. Most Darwin disciples assume that these complex processes and micromachines arise from the gradual accumulation of all the interacting parts. Of course, pure naturalistic Darwinism precludes any goal or intentionality in any organism's evolutionary history. Each part of a "future" machine is generated by random mutation with no teleological forethought about how handy it would be to evolve this new system. Each new part is added the same way—mutation and subsequent expression of the gene until all the parts are made and present. Then all the parts happily start interacting with each other, and *voilà*, the system is up and running and giving the lucky creature an evolutionary edge over those sorry relatives that didn't receive all those chance mutations which produced all those seemingly useless parts over so many generations.

There is another option that strains one's ability to suspend disbelief. It is as follows: the creature underwent a series of macro-mutations that generated all the parts all at once, and then all the parts interacted to make a functioning system. This is comparable to a series of geological events generating all the parts of a car all at once. Then they self-assemble with earthquakes throwing the parts together.

Finally, the finished car rolls out from a new steaming fissure in the side of a mountain. The creationist simply explains these fantastic micro-machines as products of a Creator Who designed, created, and integrated these amazing systems in their entirety within the body of a certain creature supernaturally. Evolutionists don't like this because it seems too simple. They opt for an elaborate series of chance events to explain the genesis of biological machines that defy the ingenuity of our best mechanical engineers. Why? They don't want to deal with the Master Engineer.

I mentioned various cell activities by name at the beginning of this chapter but I have not yet demonstrated how Behe's irreducible complexity thesis applies to any of them. Now I would like to describe one glorious biological micro-machine in a fair bit of detail: the contraction system of a muscle cell. I didn't list it in my fly-by overview of general cell functions since it is a system unique to muscle tissue or unicellular critters that can quickly change shape.

Before I get into the nitty-gritty of muscle contraction, I want to describe the context of this system so that there is not a mental disconnect between the system I describe and the real world muscles you use every second of every day. Let us focus on a well-known muscle: the biceps brachii. This upper arm muscle is attached to your shoulder blade on the upper end and one of the forearm bones (radius) on the lower end. Muscles only do two basic things mechanically: contract and relax. When this muscle contracts it pulls (flexes) the forearm toward the biceps (Figure 26).

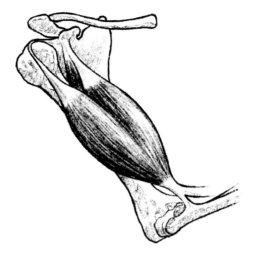

Figure 26: Biceps brachii.

The biceps, like all skeletal muscles, are composed of a parallel arrangement of small bundles of muscle tissue called *fascicles*. These may be familiar to you if you have eaten a well-cooked pot roast. The roast fascicles can be easily teased apart with a fork. Each fascicle is more or less the diameter of a coffee straw but not as round. A fascicle is a bundle of hundreds of parallel muscle cells (muscle fibers). We are now into the microscopic realm. Each muscle cell is exceedingly thin (approximately the thickness of one strand of hair) but may be relatively long, about an inch (more or less). If you could enlarge one muscle cell so that its diameter was the width of a package of spaghetti, you would find it to be many feet long. A spaghetti package represents a very short section of a muscle cell. The plastic wrapping corresponds to the cell membrane or boundary of the cell. Within the muscle cell are hundreds of contractile cylinders called *myofibrils* (Figure 27).

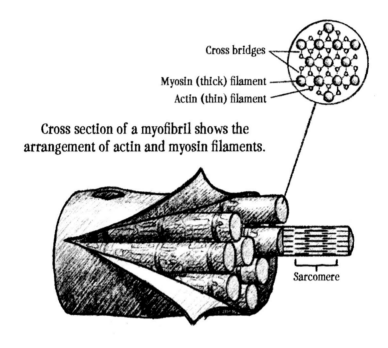

Figure 27: Section of a muscle cell containing myofibrils.

These are represented by the parallel spaghetti noodles in the package. This model is only useful to describe the spatial arrangement of the contracting machinery in the cell. One myofibril (noodle) is composed of many sarcomeres linked end to end like boxcars in a train. Each sarcomere is the fundamental contracting unit of the cell.

Are you with me so far? Pardon me if I mix my metaphors, but if every boxcar in a train suddenly shortens, then the train would shorten. In the same way, when all the sarcomeres shorten, then the myofibril shortens. And when all the myofibrils in the cell shorten, then the muscle cell shortens. When all the muscle cells in a fascicle shorten, then the fascicle shortens. And lastly, when all the fascicles in the muscle shorten, the whole muscle shortens! Now, let us go back to the sarcomere because we're not done yet. We are now down to the protein level, a level that is hard to imagine three-dimensionally. The sarcomere is composed of several parts (mostly proteins) that are essential in the contraction process. They are listed below:

1. myosin (protein)—thick filaments
2. actin (protein)—thin filaments
3. tropomyosin (protein)
4. troponin (protein)
5. sarcoplasmic reticulum or SR (organelle that stores and releases calcium)
6. T-tubules (extension of the cell membrane that transmits the impulse to the sarcoplasmic reticulum)
7. ATP (energy source)

Figure 28 shows the arrangement of these parts in and around one sarcomere (I didn't list the Z disc in an effort to simplify everything, but it is required too).

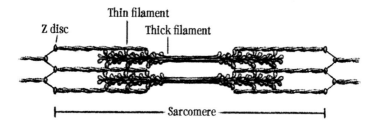

Figure 28: A sarcomere.

How do these parts interact so that the end result is the shortening of the sarcomere? First, hold your hands with three pens held by your eight fingertips as illustrated in Figure 29.

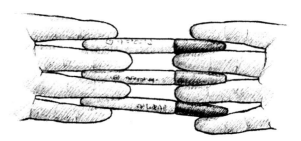

Figure 29: Fingers and pens representing a sarcomere.

This is a very crude illustration of a sarcomere in a relaxed position. The base of your fingers represent the ends of the sarcomere. Adjacent sarcomeres to your right and left would be attached to both of these ends. Now carefully slide your two sets of fingers toward each other until your fingertips touch and the pens slide to the base of your fingers. This illustrates a sarcomere contracting.

This gives you the basic idea of a shortening sarcomere, but it does not explain the mechanism driving the movement. The fingers and pens will serve to introduce two of the proteins. Your fingers represent the *actin* (thin filaments) and the pens represent the *myosin* (thick filaments). When you slide your fingers together, the pens and fingers slide past each other. What is actually happening within the sarcomere? The myosin proteins (pens) have many oar-like extensions called myosin heads (I like to refer to them as myosin oars because they act more like oars) that attach to myosin-binding sites on the actin proteins (fingers) like oars pulling the actin (fingers) toward each other. Now I must dispense with the fingers and pens to illustrate in more detail how contraction and relaxation can occur by nervous control. Remember the *sarcomeres* are linked end-to-end to form a long cylindrical *myofibril*, and many parallel myofibrils fill the innards of the muscle cell. In addition to all this, a membrane-bound organelle called the *sarcoplasmic reticulum* is sandwiched in between all these myofibrils. This organelle is a storage container for *calcium* and when it receives the appropriate signal (a nerve impulse) it releases calcium into the adjacent sarcomere where it has a profound effect. But first I must introduce two more proteins, both of which are wrapped around the actin (thin filaments). Remember that the actin (fingers) is the protein on which the myosin oars attach and pull, shortening each sarcomere. Of course if the myosin oars were constantly tugging on the actin in all the sarcomeres in all your muscles, you would be in a very unhappy state of tetanus or uncontrolled contraction until death. In order to have any meaningful movement you must have the ability to contract or relax muscle fibers at will. This is where the calcium, troponin, and tropomyosin come in. *Tropomyosin* is coiled around the actin thin filaments like strings coiled around my fingers. When the sarcomere is in a relaxed state, this protein physically blocks the myosin oars from attaching and pulling on the actin. What role does the troponin play? The tropomyosin doesn't have a mind of its own nor is it directly controlled by nervous stimulation so *troponin* acts like a switch. When troponin changes in its 3D shape, it literally levers tropomyosin into a different position exposing the myosin-binding sites for the myosin oars. When the binding sites are open the myosin oars attach, power stroke, release, recovery stroke, attach, power stroke, release, recovery stroke, attach, power stroke, and so on until the sarcomere cannot shorten any more. If you are following me so far, then you might think troponin acts as a switch determining if there will be a contraction or not. If so, you would be correct. So what flips the switch? *Calcium* is the culprit. As I said earlier, calcium is released from the sarcoplasmic reticulum in response to the nerve impulse. Calcium immediately binds to troponin causing it to change shape. Troponin's change results in tropomyosin being levered out of the way. The myosin oars are then free to stroke, stroke, stroke. One other structure in the cell helps this process along, but it might not be required. The *T-tubules* are indentations of the muscle cell membrane that form a subway system into the interior of the cell and interface with the sarcoplasmic reticulum. Essentially the T-tubules relay the nerve impulse from the cell surface (where the neuron delivered the impulse) down into the cell where it can more efficiently stimulate the sarcoplasmic reticulum. Let's summarize the process and then explore its irreducibly complex nature.

1. A nerve impulse arrives at the muscle cell (by the way, the nerve impulse is, in itself, another irreducible complexity).

2. The impulse is transmitted down the T-tubules into the cell's innards.

3. This impulse stimulates the sarcoplasmic reticulum to release its stored calcium.

4. The calcium binds to the troponin which causes it to change shape.

5. Troponin's change of shape causes it to lever tropomyosin out of a blocking position.

6. The myosin oars attach to the myosin-binding sites, and…

7. stroke…stroke…stroke.

You might be wondering where the ATP comes in. Just like any work, energy is expended. In this case, the energy stored in the ATP is used in the recovery stroke. To get really specific, ATP attaches to the tip of the myosin oar, causing it to detach from the actin. This is kind of like lifting the oar out of the water. The released energy from the ATP causes the myosin oar to be re-cocked (recovery stroke) so that it is in position for another power stroke.

MISSING PARTS

Now I would like to establish this nifty nano-machine as an irreducible complexity. It is quite obvious that if you didn't have actin, there would be no surface on which the myosin oars could attach and pull. Result: no contraction. What if myosin was missing? Then there would be no oars to attach to the actin. Result: no contraction. What if you didn't have the tropomyosin? This gets ugly, but if everything else was in good order, you have no way to block the myosin-binding sites. If ATP was available, the oars would keep on stroking and stroking. In short, you would be in a constant state of tetanus. If tropomyosin was in its blocking position (relaxed state), it would stay that way if troponin was missing. Troponin is the protein switch that moves tropomyosin out of the way or in the way of the myosin oars. If there is no troponin, there would be no way to toggle back and forth between open and closed sites for myosin to perform its series of power strokes. Outcome: contraction or no contraction depending on the position of tropomyosin. Not to belabor this, but what about no calcium? Calcium is the finger that flips the troponin switch, or to be technical, it causes the troponin to change shape which moves tropomyosin out of the way of the myosin-binding sites. Outcome: no contraction possible. How about no sarcoplasmic reticulum? If this is the case, then there is no container to store the calcium. If calcium is loose in the cell fluid, then the 'switch' troponin would be in a permanent 'on' position of holding tropomyosin out of the way of the myosin oars. Again the outcome would be a perpetual contraction (tetanus). T-tubules are involved in delivering the impulse directly to the sarcoplasmic reticulum but it may be possible for the impulse arriving at the cell membrane to stimulate the sarcoplasmic reticulum without the go-between of the T-tubules. Regardless, even if we eliminate the T-tubules as an obligatory component, this contraction system is an irreducible micro-machine. If you talk to a muscle cell biologist, you would find that I left out all sorts of cool details. I daresay there are many more parts to the system to make it work smoothly in the context of a living animal. I have presented the bare essentials needed for the contraction to occur. I may have even missed some important details, but the point remains the same: muscle micro-machines cannot evolve a step at a time. There are many other things that I have not discussed which would make a Darwinian explanation even more laughable. For instance:

1. Feedback mechanisms to make the muscles grow when worked.

2. Developmental regulation to orchestrate muscle growth so that each muscle attaches to the correct bones (or other structures) in the correct places to produce useful movement.

3. Making enough individual muscles throughout our body to ensure the wide range of motion we have.

4. Gene regulation to ensure that the correct genes are turned on so that these muscle proteins are made in the right quantities and assemble themselves in the right way.

5. Controlling the amount and timing of nerve impulses needed to enlist the appropriate number of muscle cells in proportion to the load (so you don't punch yourself in the face when you decide to scratch your nose).

6. Other factors that enable this system to be integrated with the nervous system, circulatory system, lymphatic system, etc.

HOW DO DARWINISTS EXPLAIN ALL THIS AND MORE?

Not well. It's not because they're dumb; it's because they mustn't. It's their moral imperative. If you understand Darwinism, it is essential that the system can never be the goal. Imagine the first organism that evolved the very first contractile apparatus (probably a protozoan like an amoeba). All the parts (usually proteins) must be encoded on the DNA. Presumably they didn't have these genes, so in order to get these genes, extra copies of other genes were available to accumulate genetic mistakes (mutations). These mutations were not directed by outside intelligence. Eventually a mutated gene was turned on and one of the proteins needed for the contractile apparatus emerged. Remember that the gene (DNA coding for a protein) specifies the sequence of amino acids.[31] The sequence of amino acids constrains the 3D shape of the protein and the

---

31 Genes can have other functions that boggle the mind. I'm sticking with the basics of protein-coding genes so my reader won't get overwhelmed.

shape determines its function. But this lonely protein can't produce movement apart from all the other components. If it is to be retained in the cell, it must earn its keep by itself. It must perform some other useful function for the cell before the future contractile function is realized. If it doesn't have a selective advantage, the cell will stop making the protein (if it doesn't stop making a useless protein, the organism's health may be seriously affected). This scenario must happen repeatedly until all the proteins are made and intermingle with each other to generate this highly sophisticated contractile apparatus.

GOOD LUCK

Here are some conditions that must be met by this newly-evolved, irreducibly complex system.

1. All the essential parts are manufactured in the right place, at the right time, and in the right quantities.

2. All essential parts (actin, myosin, tropomyosin, etc.) must be pre-adapted to function together. For example, auto parts designed for a Ford F-150 can't be replaced by similar auto parts designed for a Honda Pilot. Auto parts that work together in an integrated system are designed to fit each other and function together. You also can't take randomly generated parts from a machine shop, throw them together in a drier, run it on tumble, and expect to get a functional machine when you open the door.

3. All essential parts either self-assemble or have an assembling mechanism that puts it all together so that it works right. This assembling mechanism cannot be intelligently designed. Have you ever wished that gently shaking the box that contains your unassembled $2,000 mountain bike would assemble it? Even though you know that all the parts are designed by intelligent engineers to fit together, you also know that it will take further intelligence and energy to read the instructions and assemble it yourself.

4. Once assembled, the system works right and doesn't interfere with other systems that are still needed. If randomly accumulating parts in the kitchen self-assembled into a mountain bike which is then ridden around the kitchen by the household teenager, it would most certainly not improve the process of cooking.

5. The system must be compatible with the energy source (ATP). If randomly accumulating radio parts self-assembled to produce a radio that takes AA batteries, it won't work if you don't have a proper outlet or AA batteries. Sticking the little plug into a 220-volt outlet or forcing C batteries into the little AA battery chamber isn't going to cut it.

6. Once assembled, the system must confer a selective advantage to the organism in its struggle for existence.

THAT TROUBLESOME INFORMATION THEORY

When we examine any naturalistic scenario for producing irreducible complexities in light of Information Theory, it is simply untenable. One aspect of Information Theory is that intelligence is required to generate information. This theory is based on countless empirical observations. Because both the proteins of this contractile system and the assembly machinery are encoded in the genes, it is reasonable to assume that the origin of this genetic information is intelligent. It is also *un*reasonable to assume otherwise, because it contradicts all evidence to date demonstrating that information must come from an intelligent source. The only way they can try to avoid this problem is to insist that genes are not information. By the way, the study of gene sequences is called *bioinformatics*. So here's the conclusion of the matter.

According to Information Theory, unintelligent, undirected mutations cannot produce genetic information encoding all the parts and mechanisms essential for the assembly and operation of an irreducibly complex system. According to all the *evidence*, unintelligent, unguided mutations have never produced genetic information encoding all the parts and mechanisms essential for the assembly and operation of an irreducibly complex system of even the most modest kind.

To God be the glory. Darwinism be damned.

CHAPTER 9
Faded Genes—Does Genetic Information Erode?

In Chapter 4: "Designer Genes," my objective was to clearly articulate that *all* genetic information is designed by intelligence. This necessitated a discussion showing the difference between microevolution and macroevolution. It is easy by Darwinian sleight of hand to obscure the fact that more complex organisms require new and different genetic instructions to be added to the genome. The available toolkit of Darwinian evolution is utterly incompetent at producing brand new information through typographical errors (mutation) to the DNA regardless of how much time and natural selection is added. In short, mutation, selection, and time can't build new genes specifying new body parts or body plans. But the situation is actually much worse. In this chapter, we'll find that mutation and time not only fail to qualify as an asset, they are actually a liability to the genome.

The second law of thermodynamics states that every system tends toward more and more disorder and less useful energy (e.g., work degrades to heat, and heat dissipates into the environment). The measure of disorder is termed *entropy*. In everyday life, we are very familiar with entropy. Everything we own that is subject to use falls apart, wears out, breaks down, or dies—our body, clothing, shoes, lawn mower, car, computer, carpets, dish towels, etc. In our current universe, *anything* used over time doesn't get newer and better and more complex; it gets older and worse and less complex. Nothing (made of matter and energy, anyway) is immune to entropy, including our precious DNA. In other words, our genes don't get better with use. Like our blue jeans, our DNA genes fade with wear. This chapter is meant to bottom-line the most salient points of an excellent book written by Dr. John Sanford, called *Genetic Entropy*.[32] If you want a more detailed, thoroughly referenced account of the inexorable degradation of our genome, read it. If you don't want to, then this chapter may give you the main thrust of the book in simpler language.

Evolution in the sense of simpler organisms evolving into more complex organisms requires a gain of information. Mutation sifted by natural selection is believed to add more information to genomes. In Chapter 4, we found out that this is science fiction. In this chapter, we find that it actually *subtracts* from the information that is already there. If I opened this document on my computer and started writing but every so often I blindfolded myself and randomly shot a spit wad at my keyboard, it may add letters and numbers and spaces and punctuation or delete them if the spit wad hit the backspace key. Regardless, the end

---

[32] John C. Sanford, *Genetic Entropy* (Canandaigua, NY: FMS Publications, 2014).

result would not be publishable. These typos would never add any sensible sentences that are in keeping with the theme of this book. If I sent it to the publisher after doing this without attempting to correct them, I should expect sarcastic comments as to my proofing abilities. The point is, characters may have been added but information wasn't; rather it was lost. If the editor didn't fix them, the manuscript wouldn't be fit to publish. But this kind of thing is the only hope Darwinian evolution has to alter or add to the genome. They rely on natural selection to eliminate those typos that subtract from the value of the book and they rely on natural selection to keep those typos that add more meaning and value to the book. Good luck! Intuitively we know that an occasional typo from a spit wad may not utterly ruin the book because it may have *alttered* a word such that the meaning was retained. But if it occurred to a word that changed the *meating* or changed it into *globberish*, then we have a problem. But one thing is for sure; it won't add meaningful chapters, let alone turn Fox in Socks into Anna Karenina.

    Darwinists say that natural selection will come to the rescue and save the organism from the degradation of the genome. Unfortunately (for the Darwinist), that doesn't happen except for the worst mutations. For example, if a mutation was really bad and caused the death of the organism or adversely affected its fitness, then natural selection would successfully eliminate it either quickly or slowly. But this doesn't happen for most mutations. Most mutations are seemingly neutral. Nothing bad happens. The mutation essentially flies under the radar. It's like a xerox copy of the original document. It is never as crisp and clean as the original, but the slight imperfections that creep in don't affect its readability. These mutations are technically not completely neutral; they are near neutral. This means that they are slightly harmful, but not enough to be selected against.

Let's say there was a slight error (near-neutral mutation) in the assembly instructions of a Buzz Lightyear in a toy factory. The factory assembly line built Buzz according to these slightly-off instructions but the error was so minor that it was invisible to the Buzz inspector and consequently passed inspection (i.e., natural selection didn't eliminate it). Next, it was approved for shipment to the retailer and bought by some child who enjoyed Buzz for almost as long as other kids enjoyed perfect Buzzes. This is an illustration of a near-neutral mutation. In real life, many of these types of mutations are not successfully eliminated. The mutated organism is able to reach adulthood, lead a healthy life, have offspring, and eventually die of causes not related to its mutation. Consequently the mutations are passed on to the next generation. At first blush this doesn't seem to be a problem. If the parent suffered no ill effects from it, then why should the offspring? But here's the rub. The offspring will add its own mutations to the seemingly harmless mutations inherited from the parent. Each generation does this and the mutations inexorably accumulate. It's analogous to a xerox of a xerox of a xerox…repeated hundreds of times. The clarity of the descendant copies will get progressively worse. Eventually this generational accumulation of minor errors will add up to a major mutational load that is not sustainable. Ultimately it results in extinction. Natural selection slows this process down but it can't stop it. Why? If natural selection eliminated the slightest error, then everything would be eliminated. Natural selection can't delete typos like a book editor and leave all the rest; it either deletes the entire organism or lets it live.

CHAPTER 10
According to Their Kinds

SETTING THE STAGE

In light of the fact that the natural processes discussed in Chapter 4 don't produce designer genes (novel innovations), it follows that there are clear limits to change. This means that the notion of the grand evolutionary tree is false. If you start with single-celled bacteria you will end with…singled-celled bacteria. In order to evolve into new, more complex life forms, new information must be infused. Even if we grant the complete Darwinian toolkit of mutation, natural selection, genetic recombination, gene flow, genetic drift, etc., tinkering away over 3.8 billion years, it isn't remotely up to the task of generating anything even a little bit more complicated than themselves. But we look around and we see a world teeming with millions of species spanning bacteria, protozoans, fungi, plants, and animals. Complexities that boggle the mind abound in all of them. These complexities may be shared with many other life forms and some are quite unique to a single group. This raises the question, if they didn't evolve naturally from one common ancestor, how did they come to be? As is often the case for big questions, go to Genesis for the answers. The answer to that particular question is found in Genesis 1:11–13 which says,

> And God said, "Let the earth sprout vegetation, plants yielding seed, and fruit trees bearing fruit in which is their seed, each **according to its kind**, on the earth." And it was so. The earth brought forth vegetation, plants yielding seed **according to their own kinds**, and trees bearing fruit in which is their seed, each **according to its kind**. And God saw that it was good. And there was evening and there was morning, the third day (emphasis added).

In verses 20–21, it says,

> God said, "Let the waters swarm with swarms of living creatures, and let birds fly above the earth across the expanse of the heavens." So God created the great sea creatures and every living creature that moves, with which the waters swarm, **according to their kinds**, and every winged bird **according to its kind**. And God saw that it was good (emphasis added).

And verses 24–25 and 27 say,

> God said, "Let the earth bring forth living creatures **according to their kinds**—livestock and creeping things and beasts of the earth **according to their kinds**." And it was so. And God made the beasts of the earth **according to their kinds** and the livestock **according to their kinds**, and everything that creeps on the ground **according to its kind**. And God saw that it was good…So God created man in his own image, in the image of God he created him; male and female he created them (emphasis added).

In this great passage, we see that God created the various plants and animals according to their kinds. As much as I would like to be able to point to particular groups of plants and animals and confidently assert that they are created kinds, unfortunately, I can't. However, we can, to some extent, say what created kinds are not. From scriptural authority we can confidently say it isn't "all of life." In other words, a created kind isn't one universal common ancestor from which all life forms, past and present, descended through natural or supernatural processes. A straight-forward reading of the text shows that theistic evolutionists or progressive creationists don't have a leg (or fin that evolved into a leg) to stand on. If it isn't one universal common ancestor, what is it? That is the million dollar question. At the risk of laying too much groundwork, I think it will be helpful to start at square one and see what can be answered from Genesis regarding what does or doesn't constitute a "kind." I will assume you are familiar with the Linnaean levels of taxonomy, i.e., kingdom, phylum, class, order, family, genus, and species. Let's start at the most general, which is kingdom (I won't bother with domain since it would require more explanation) and eliminate possibilities other than one universal common ancestor. Since the Bible doesn't speak of protists, fungi, and bacteria, we will limit our discussion to plants (Kingdom Plantae) and animals (Kingdom Animalia). God did not create one common ancestor of all plants because the text says "plants yielding seed *according to their own kinds.*" God did not create a common ancestor of all animals either because the text says *birds, sea creatures* (day 5), *beasts, livestock, and creeping things* (day 6) *according to their kinds*. Also, the last phrase "according to their kind" isn't implying that birds, sea creatures, etc., are the "kinds." For example, when the text says, *"and every winged bird according to its kind"* it is clearly indicating that there is a variety of winged birds from the start. "Every winged bird" doesn't mean one kind of winged bird. This is also true for the other general groups like plants, sea creatures, beasts, and creeping things. From the plain reading of the text, there are a bunch of different kinds of plants, sea creatures, beasts, cattle, and creeping things. The Hebrew words for beasts, livestock, and creeping things have no specific correspondence to a particular taxon like class, order, or family. But whatever the grouping, it indicates that there are different kinds of that grouping. Birds don't mean exclusively Class Aves (birds). It means all flying creatures which would include birds, bats, and pterosaurs. So the scripture doesn't give us much detail below class (classes of land vertebrates are amphibians, reptiles, birds, and mammals). The Hebrew word *behemah* translated "livestock" or "cattle" denotes domesticated beasts of burden (presumably cattle, camels, goats, sheep, etc.). Creeping things (*remes*) isn't very specific either. It generally means reptiles, amphibians, insects and other assorted creepy crawlies, including other invertebrates (slugs, spiders, scorpions, worms, etc.). As is obvious with this array of creatures, *remes* isn't any more precise than our phrase "creepy crawlies." Therefore, we need to think it through using Scripture and observational science to ascertain what a created kind actually is.

DO WE REALLY NEED TO KNOW?

Since Scripture isn't clear as to what constitutes a kind, why bother trying to find out? Good question. Secular evolutionists believe that all life shares a common ancestor (one enormous tree of life where all branches have *continuity* with one trunk) and creationists don't believe that. We believe according to the scripture quoted above that God created life having diversity from the start with many separate trees (created kinds *with discontinuity* between trunks) each branching into several to many twigs (species). So an evolutionist may reasonably inquire, "If you guys don't believe in one big evolutionary tree with one common ancestor, then you must believe that God created a bunch of little trees (created kinds) each with a capacity to branch, right?" And we answer, "Yep! We just can't figure out how many 'trees' or how big each tree is." It isn't shameful that we don't know but it certainly is to our glory to try and find out. Proverbs 25:2 says, "It is the glory of God to conceal things, but the glory of kings is to search things out." If the answer is attainable through the light of Scripture and science, we should do our utmost to search things out and get the right answer. It isn't enough to know they have the wrong answer.

IT ISN'T AS SIMPLE AS IT MAY APPEAR

When God made life, He didn't create everything equally different from each other. He created the entire array of life (millions of species) in a nested hierarchy. In other words, many kinds of organisms are different from each other to varying degrees than they are to other kinds of organisms. This pattern reveals itself in the Linnaean hierarchy. Some groupings, like genera, show little difference between members; others, like orders, show big differences. Kingdoms show enormous differences. For example, all box turtles are in the same genus (*Terrapene*). The differences among the box turtles species are fairly minor. Most people would say, "What differences?" However, when we move up a level to Family Emydidae (a group that is comprised of box turtles, painted turtles, sliders, cooters, diamondback terrapins, map turtles, wood turtles, etc.), they show remarkable differences that are discernible to most people with eyes in their head. In spite of their differences, however, they have more in common with each other than other turtle families like softshell turtles (Family Trionychidae) or snapping turtles (Family Chelydridae). The next level up is *all* turtles and tortoises: Order Testudines. Even though it is a big diverse order having over 360 species grouped into 14 families, all turtles share a hallmark characteristic—the shell which includes the upper carapace and lower plastron. These and other features warrant grouping them together as turtles. Though the Order Testudines (turtles) is not one created kind, it is clearly distinguished (even by young children) from all other creatures. So which level constitutes a biblical kind? Species? Genus? Family? Order? Did Noah bring on the Ark all species of box turtles or did he bring on the ark the ancestral pair of box turtles that then diversified into all box turtle species? Or did he bring on the Ark the ancestral pair that gave rise to the Family

Emydidae? The problem is that different creationists have different opinions on what constitutes a kind and so far haven't agreed on the criteria to help us answer this important question. I used box turtles (genus), Emydidae (family), and turtles (order) to illustrate the problem because I'm partial to turtles,[33] but I could have used any animal or plant to show these different groupings based on varying degrees of similarity and differences.

PARADIGM SHIFT FROM LINNAEUS TO DARWIN

Back in the eighteenth century, Carolus Linnaeus was the father of modern taxonomy and was the first to propose some of the groupings we are familiar with—kingdom, class, order, genus, and species (phylum and family were added later). He was a century before Charles Darwin and was a card-carrying creationist. During his time it was thought that all species, even very closely related species, were created separately and distinctly from each other and have continued to the present remaining essentially unchanged since the creation. This idea is called "fixity of species" and can be visually represented by a diagram called the "Linnaean Lawn"[34] (Figure 30).

---

33 I did my doctoral research on the reproductive ecology of the eastern box turtle and am extremely fond of them.

34 This term Linnaean Lawn was not coined until recently but it still visually represents the idea of fixity of species quite well.

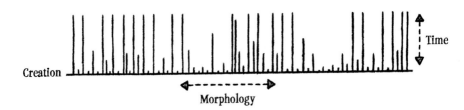

Figure 30: Linnaean Lawn.

Each vertical line (blades of grass in the lawn) represents a created species. Going from the bottom to the top is the passage of time. The top of the diagram is the present; the bottom is the creation week. A line that extends from bottom to top has survived to the present. If not, it has gone extinct. Parallel lines close to each other were created morphologically similar to each other (same genus). The lines further apart represent species less similar to each other.

But then Darwin published *The Origin of Species* in 1859 in which he essentially proposed a radical shift away from the Linnaean Lawn or the "fixity of species" paradigm. His theory of natural selection provided an explanation of how two or more similar species were not specially created distinct from each other. Rather, they arose from a recent common ancestor. Say an original population was split by a geographical barrier like a mountain range or large river. The separated populations became increasingly different from each other through natural selection slowly adapting them to different sets of environmental conditions. Eventually they became different enough physically and behaviorally so they no longer would interbreed with each other, even if the geographic barrier was removed. At this point speciation (one species becoming two or more species) has occurred. But Darwin doesn't stop there. He proposed that common ancestors of each genus shared a common ancestor for the entire family and the common ancestors of each family shared a common ancestor for the entire order... and so on down to one common ancestor (or several) of all living things. In other words, he converts the Linnaean Lawn into an evolutionary tree (Figure 31).

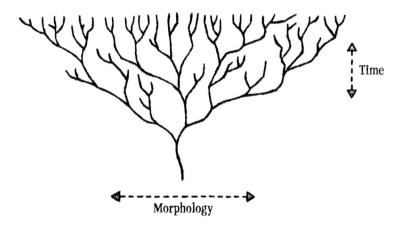

Figure 31: Evolutionary Tree.

But even if we grant speciation (which most creationists do), the problem of generating novel specified information in one or more of the descendant populations still stands. Natural selection is all well and good, but it doesn't create novel genetic information even if branching (speciation) occurs. This is what I focused on in Chapter 4. Therefore the evolutionary tree is wrong not only on biblical grounds but also on scientific grounds. So branching or speciation *isn't the problem*. Random mutation generating complex specified genetic information that makes descendant species more complex than their ancestor *is the problem*. This is the crux of the matter and should guide us in trying to determine the actual created kinds.

BARAMINS AND BARAMINOLOGY
(A MODERN CREATIONIST APPROACH)

For obvious reasons, creationists reject the evolutionary tree but do we really need to go back to the Linnaean Lawn? No. There is an alternative: the creationist orchard. Creationists understand that the created kind could undergo *limited* change over time. This change does not exhibit any increase in complexity or net gains in novel genetic information; it simply means that God created the various kinds with the genetic potential and versatility to adapt to an ever-changing environment. These created kinds (also known as *baramins*, from the Hebrew words *bara*, "to create" and *min*, "kind") had the God-given capacity to speciate (adapt and diversify) into a number of similar species. So each *baramin* or *created kind* is a little metaphorical tree. God created the various "kinds," and after the flood, these kinds were able to diversify (within limits) into a number of species. So each becomes a "little tree" with all the twigs being the species of that baramin. Since there are many different kinds, we have a metaphorical "creationist orchard" (Figure 32).

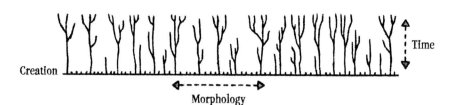

Figure 32: The Creationist Orchard.

It is easy in principle to agree on the concept of a branching tree within the orchard, but disagreements may arise when we start delineating a single tree. These baramins may correspond to certain groups known today, such as a genus or even a family. For example, here is a box turtle tree from a secular source (Figure 33) that assumes that this is a little branch on the grand evolutionary tree.

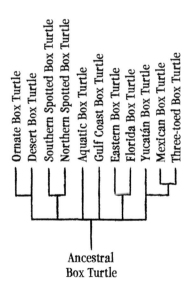

Figure 33: Box Turtle Tree.

However, a creationist could easily accept it being one tree in the creationist orchard. But given the creationist orchard paradigm, some creationists think this tree is too small and that it needs to include the entire Family Emydidae like so. The box turtles in the genus *Terrapene* are circled in the middle of this tree (Figure 34).

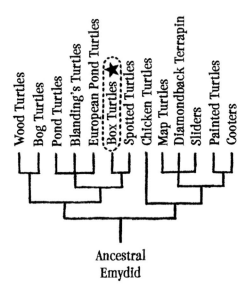

Figure 34: Family Emydidae Tree.

See the dilemma?

Although my personal bias is toward smaller, conservative baramins, my desire is not to tell you *what* to think on this issue but rather *how* to think through this issue. So without coming to any firm conclusions, I want to give you various criteria that we all should consider when defining a particular baramin or assessing a baramin already defined.

Creationists over the last century have wisely taken into account the ability of species to diversify through various mechanisms including phenotypic plasticity,[35] mutation, natural selection, genetic recombination, etc. Although research time frames make it unlikely to observe a speciation event, most creationists assume speciation can (and does) happen. However, we are still faced with the question of "How much and to what extent can speciation happen?" In other words, "How big and diverse can a baramin be?" My goal is not to tell you how big or small, how limited or diverse a baramin is, but rather I want to mention some things baraminologists should consider in trying to delineate a created kind. As you might guess, baraminology is the study of baramins. It is a fascinating yet troublesome topic. It is fascinating because we creationists should strive for a good answer for evolutionists who insist on one evolutionary tree (continuity) when we insist on separate trees (discontinuity) in the orchard. It is troublesome because baraminologists are not in agreement on the criteria to delineate baramins. It is also multidisciplinary. Genetics, hybridization studies, biochemistry, paleontology, biogeography, morphology, etc., are all areas that can speak to the topic of baraminology. Multidisciplinary topics will often attract an assortment of scientists (of different specialties, strengths, weaknesses, blind spots, and assumptions) to the table.

So what should we consider? It is very easy to let "the experts" give us the answers, but even lay Christians interested in creation science need to be tentative about the

---

[35] Phenotypic plasticity is the ability of an organism to change in response to stimuli or inputs from the environment.(M. J. West-Eberhard, in *Encyclopedia of Ecology*, 2008).

conclusions of the experts. Experts also need to be tentative about their own conclusions. Remember we've heard time and again that the past cannot be repeated, tested, or observed. What the *created kinds* actually were isn't empirical science. It is historical science which means we need to be exceedingly careful about our assumptions and criteria, and be tentative about our conclusions.

CONSIDERING THE CRITERIA

Successful Hybridization

If two organisms can successfully breed and have offspring (fertile or sterile), it was assumed that they belonged to the same kind. I always thought this was the most robust criterion that the organisms belonged in the same kind until I realized that the Bible doesn't necessarily say that it is impossible for different kinds to breed. In other words, this is not a slam dunk criterion.

Similar Morphology

If two species are very similar morphologically, it may be one indicator that they may be in the same baramin. For example, all cats (Family Felidae) have a cat morphology. Even a toddler knows the cat gestalt. At the zoo she might point at a huge Siberian tiger and say "Kitty!" (There are many creatures that, to the novice, look similar morphologically at a mid to high taxonomic level, like family or order, but there may be no other evidence that they share a common ancestor. To the specialist they may be worlds apart behaviorally, anatomically, ecologically, and genetically. They don't mate nor are they inclined to do so, or, if they do, don't produce fertile or infertile offspring.) Our initial assumption should be that they belong in separate baramins until proven otherwise, rather than assume they belong in the same baramin until proven to be separate. This is a very weak criterion because two organisms can be identical genetically but have vastly different appearances simply because their genes are being expressed very differently. Conversely, they might be quite different genetically but have a very similar morphology. In summary, morphology is too variable based on differential gene expression to be a clear indicator. Also, similarities based on common design (at mid-taxonomic levels) can also "muddy the waters" of baraminology.

Genomic Equivalence

I propose that differences in genetic information content is the most consistent measuring stick to determine baramins. The reason for moving away from morphology, as mentioned before, is that differential gene expression can result in mildly to wildly different morphologies in genetically identical organisms. Consider complete metamorphosis in insects. The same individual caterpillar turns into a butterfly or a maggot turns into a fly. The morphologies of larva and adult are so radically different they could legitimately be considered in separate phyla if we didn't know they were the same species. In fact, William M. Wheeler (1908) noted that an entomologist named von Heyden, when first discovering an unknown larval form of a hoverfly that lived in ant colonies (Family Syrphidae), thought it was a mollusk! **The goal of a baraminologist should be to distinguish (and this is the tricky part) between the genetic similarities due to common descent and those due to common design.** Within one real baramin, the genetic similarity should be at or near 100%, which also means that it is most likely due to common descent.

All creationists reject the notion that humans and apes share a common ancestor and rightly so. When faced with the evolutionist's claim that we are related to apes, we amass all the ecological, morphological, and genetic evidence to show that there are significant differences between all the apes and humans, differences that show we are distinct. We acknowledge that we may share genes A–W with chimpanzees, but we are quick to point out that we have genes X, Y, and Z that apes do not. After conclusively separating humans from apes using observational science (comparative anatomy, genome comparisons, etc.), we then lump together into one baramin members of one family of animals or plants that have the same if not a greater degree of ecological, morphological, and genetic disparity. All these differences are important to consider but the latter, I think, is the most fruitful as we reconsider the boundaries of baramins.

The Genomic Equivalence Criterion—A Better Measuring Stick

Because naturalistic processes cannot generate new genetic information, it would be prudent to consider this when looking at any group under consideration for baramin status. Hybrid data or not, we must ascertain whether or not the genomes of all these species are essentially equivalent. What I mean is this. There needs to be close to 100% equivalence of the whole genome (all coding and non-coding DNA) in all species being considered in order to tentatively conclude they are within the same kind. The standard of genomic equivalence should be calibrated by the amount of difference exhibited in the human genome across all people groups (which is roughly 0.1%). All humans are about 99.9% the same. Why so strict? Because this is one of several ways creationists separate humans from the apes. It is because mutation and natural selection cannot generate novel genetic information. It doesn't matter how small the increase is. If species 10 has more complex, specified genetic information (the percentage is irrelevant and whether it is coding or non-coding is irrelevant) than species 1–9, then it should not be placed in the same kind with species 1–9. It is important to note that a new or different allele is not new genetic information, it is simply a variation on a gene already present. What about hybrids of species quite different from each other but within the same family like the dog-fox hybrid, the camel-llama hybrid or the horse-zebra hybrid? If genomic equivalence is a valid criterion, it behooves us to compare genomes of these organisms as well. Understandably this cannot be done overnight. Much baseline data must first be collected. This includes whole genome sequences of species to be compared (funds to do this must also be available), and research must be done to determine what genetic differences[36] can be allowed and

---

36 Allelic differences, gene duplications, ERVs, etc.

still be considered genomically equivalent. This approach cannot be too wooden. There are a few possible complicating factors that may result in falsely concluding non-equivalence (or falsely concluding equivalence):

- Chromosome comparison: If chromosome number differs, it does not necessarily indicate that the species aren't equivalent. It may be due to chromosome fusion or other chromosomal mutations. We must look at both coding and non-coding DNA to see if there are any differences in the information content of the species being considered. Genetic information could be equivalent but chromosome numbers could differ. Obviously chromosome numbers could be the same but the information content could be a little to very different.

- Endogenous retroviruses[37] (ERVs): Genomic differences due to the number, type, and placement of retroviruses are not necessarily indications of genomic non-equivalence. Conversely, similarities in the number, type, and placement of ERVs do not necessarily indicate equivalence. These sequences need to be carefully assessed. Conclusions that they are invasions of foreign DNA and/or junk DNA have been shown to be premature or simply wrong.

---

37 Any DNA segment, presumed to originate from a retrovirus, that is found in the genome of an organism's germline.

- Gene duplication/allele[38] differences: Differences in the number of copies of certain genes and differences in alleles do not necessarily indicate non-equivalence. Again, information content is the issue.

- Ploidy[39] differences: Two species may have different ploidy but may show clear genomic equivalence. The Cope's Gray Treefrog is diploid and the Gray Treefrog is tetraploid. Although they do not interbreed, they are morphologically identical (voice differs), and there is good evidence to assume genomic equivalence. This difference probably arose through unsuccessful reduction division during spermatogenesis and oogenesis resulting in diploid sperm and egg. Fertilization resulted in a tetraploid offspring. *H. versicolor* (the Gray Treefrog) has twice the amount of DNA but the information content is virtually the same.

In summary, though differential expression of a given genome can generate an enormous range of different morphologies (body shapes), the information content of a particular created kind's genome is going to be fairly stable. Here's an illustration. A deck of 52 cards can produce a great variety of hands. The genome of a particular baramin is like a deck of cards. Other baramins have genomes that have different decks with a different number of cards (some more, some less) as well as unique cards and even

---

38 Alternative version of the same gene.

39 The number of complete sets of chromosomes in a cell.

suits. Even if they share a lot of the same cards and suits, it doesn't mean they are the same kind. Different suits and unique cards put them in a different group altogether. In the same way, if a species has a certain amount of completely novel genetic content that other species don't share, it should be assumed that they don't share a common ancestor. That's one way we separate chimps from humans. **It doesn't matter how much we share; what matters is what we don't.** What we don't share is what tells us we don't have a common ancestor. Some say that a certain species may belong to a baramin even if it doesn't have genes that the rest of the group does. They say it's possible for one species within a baramin (a twig in a creationist tree) to lose genetic information that the rest of the baramin still has. With that line of reasoning, then what's to keep people from lumping chimps in with humans? We can respond by telling them we have genes X, Y and Z and chimps don't. But using the same reasoning above, they can claim that during chimp speciation, genes X, Y, and Z were lost. See the dilemma? But, complete loss is unlikely since genes stay put. "Loss" usually means mutations that degrade the gene(s) into non-coding sequences, but you should still be able to detect homology.[40] If chimps were a degraded twig in the human tree, then we should still be able to see genomic equivalence. The "lost genes" argument cuts both ways. It's a double standard if we use it on animals (or plants) to lump genetically disparate species into huge baramins but then call "foul" when the same reasoning is used to lump chimps and humans together.

---

40 Homology is a similarity often attributable to common origin (https://www.merriam-webster.com/dictionary/homology).

The Fossil Record

Another field that sheds light on where the breaks or discontinuities between baramins are is in the field of paleontology. Paleontologists can look at a particular group's (genus or family or order) record in the rocks. In other words, they can find when it appears in the older strata and when it disappears in younger strata (if it has gone extinct). The range of strata that a group occupies is called its stratigraphic range. If the fossil record of say a family of beetles stays pretty distinct throughout its stratigraphic range from where it first appears, and its characteristic morphology doesn't meld with other beetle families into what appears to be a common ancestor of all beetle families, then the fossil record informs us that discontinuity is at least between families. It doesn't at all suggest that all beetles came from a common ancestor. In Chapter 5, I graphically showed how the fossil record really highlights discontinuity, not only between phyla but also down to family level groupings. When most everything is missing except the skeletal remains, much of the distinguishing characteristics that help taxonomists separate genera and species isn't there. This doesn't necessarily mean family is the baramin; it means that we don't have enough data to resolve possible discontinuities below family. In other words, we have no way of knowing whether one fossil can breed with another fossil. We have to be content not to know. We also don't have genetic data so we must be very tentative in our speculations.

Biogeography

A basic definition of this discipline is "a science that deals with the geographical distribution of animals and plants."[41] Biogeographers try to ascertain what lives where and why. This also draws data from the fossil record which makes things even more complicated and rife with assumptions. Biogeographers try to figure out the reasons why these plants and animals are adapted the way they are and how and why they got there in the first place. This is an exceedingly iffy discipline. The secular perspective assumes the geologic time scale, slow and gradual plate tectonics, and the belief that organisms can undergo macroevolution. None of these assumptions are correct, and therefore their conclusions are wrong even though the practitioners are very intelligent. For creation science, this discipline is young and needs a completely different perspective with biblical assumptions including a relatively young Earth (roughly a bit over 6,000 years), catastrophic plate tectonics (not proven but still a very good creationist model), the fact that humans can transport animals and plants by boat to distant places, and limited biological change (microevolution) of all organisms. These starting assumptions will help build a creation model of biogeography and also inform our thinking in baraminology.

Some baraminologists believe that vipers (Family Viperidae) constitute a created kind. This family includes two subfamilies. The vipers (subfamily Viperinae) have no

---

41 "Biogeography," https://www.merriam-webster.com/dictionary/biogeography.

pit organs and are only in Europe, Africa, and Asia (13 genera; 88 species). The pit vipers (subfamily Crotalinae) have pit organs. Their 23 genera and 216 species are distributed across North and South America, Eastern Europe, Asia, and Indonesia. If we assume that they have all descended from a pair of ancestral vipers coming off the ark, then we have to assume these two lineages (the non-pit vipers and the pit vipers) formed early on. This means that the pit vipers are more related to each other than to any non-pit vipers. To say they all descended from one created kind makes their current distribution very puzzling. For example, the Russell's Viper (a non-pit viper in India) is more related to the Puff Adder (a non-pit viper in Africa) than to the Malabar Pit Viper living in India. And the Malabar Pit Viper is more related to the pit vipers of the United States, i.e. rattlesnakes and copperheads than to the Russell's Viper in India (see Figure 35 on the next page). This makes little sense if they are all descended from one pair of viper ancestors off the ark. I don't know what produced the current distribution of vipers, but it is much easier to explain a variety of disparate created kinds (genus level) that happen to migrate to the same region, than to explain why highly disparate members of a large created kind (different genera or subfamilies) are living in the same region. Assuming large created kinds leads to convoluted and complicated reasoning to explain extreme diversification and puzzling geographical patterns. We must be open to the possibility that the different genera of vipers might be different enough to warrant each being in a different created kind.

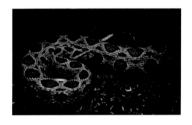
Russell's Viper (viper in India)

Puff Adder (viper in Africa)

Malabar Pit Viper (India)

Northern Pacific Rattlesnake
(pit viper in the U.S.A.)

Copperhead
(pit viper in the U.S.A.)

Figure 35: Various vipers and pit vipers.

AN UNSATISFYING CONCLUSION OF THE MATTER

Most lay creationists want clear answers to these and other foggy problems and unfortunately I don't have any. There are too many questions that are still unanswered. However, I think we have promising guidelines to use in seeking to discover created kinds or baramins. But we must be patient and not make hasty conclusions by prematurely lumping various species and genera together into larger and larger baramins before we even compare the genomes. I think our default assumption should be conservative and to start with small, modest baramins unless genetic comparisons reveal a high level of genomic equivalence at groupings larger than genus. If we prematurely assume large baramins without enough genetic similarity, evolutionists will be quick to point out that we've finally seen the light and are on the road to becoming evolutionists. The larger the group, the more cautious we need to be in labeling it a created kind. If we theorize on anything, we should bend over backwards to publicly express to the wider creationist community the tentative nature of those theories. If we lean toward large baramins (containing hundreds if not thousands of species), we also must be able to show plausible and testable mechanisms of how so many species could diversify from the "ark kinds" in such a short time. We also need to be flexible. Depending on the group, genomic equivalence may suggest large baramins but it may also suggest small. My gut assessment might be wrong on what constitutes a baramin, but it is always good to exercise critical thinking even when hearing the opinions or conclusions of the expert creationists. We must not assume that popular creationist conclusions of family level baramins are gospel truth. We should hold our pet theories about baramins in humble, open hands. Making dogmatic conclusions about baramins, with or without sufficient genetic evidence, is walking on thin ice. If we are shown to be wrong, we may unexpectedly take a chilly plunge into Lake Humiliation. Baraminology falls under historical science. We can never be certain because we weren't there when God created all the animals and plant kinds, nor did He express the boundaries of those kinds.

CHAPTER 11
Biological Badness and the Goodness of God

This chapter is about the theodicy[42] of biological "evil," namely, the presence of predators, parasites, and pathogens that handily deal out disease or death along with much pain and suffering to both mankind and animals. I will often refer to them as PPPs. Most nature documentaries are loaded with predator-prey relationships (which presumably improve ratings as most people have a morbid fascination with carnage and gore when viewed from the safety of their comfy sofa while sipping their mocha). To a Bible-believing Christian, any cursory glance at a parasitology textbook will reveal many parasites with insidious and complicated life cycles that cry out for biblical answers. This death-riddled world isn't news to anyone. We become aware at an early age that every living thing dies from one thing or another. We usually shrug it off when death is distant but it's a different story when it strikes close to home. When a loved one or a cherished pet dies, we

---

42 A defense of God's goodness and omnipotence in view of the existence of evil. "Theodicy" https://www.merriam-webster.com/dictionary/theodicy. Accessed May 2, 2023.

become acutely aware that this "bondage to corruption" was not the way things were originally created. Something has gone wrong. Of course, plant death isn't considered a consequence of the Fall, but animal and human death is. I included this chapter because Christians need a biblical theodicy of biological natural evil and ad hoc explanations won't do when unbelievers (upset by some disease or disaster) decide to lash out at some hapless Christian with their grievances against God for allowing such pain, suffering, and death. For those who read their Bibles, having a biblical answer isn't that hard. What's hard is expressing it to someone (who is either an atheist or an unbeliever angry at God) when you know your answer won't be warmly received. Regarding any calamity, caused by a biological agent or not, the answer is the same. Isaiah 45:7 says, "I form light and create darkness; I make well-being and create calamity; I am the Lord, who does all these things." Isaiah doesn't mince words here. God is the one Who creates calamity. In Amos 3:6 the prophet writes, "Is a trumpet blown in a city, and the people are not afraid? Does disaster come to a city, unless the Lord has done it?" This verse has two rhetorical questions and the implied answer is "no" to both. People are afraid when the trumpet is blown...and disaster only comes to a city when the Lord does it. These verses preclude false platitudes like, "When disaster strikes, God's heart was the first to break."

So let's begin.

The first conundrum in developing a theodicy regarding PPPs is that they don't seem to fit this verse: "And God saw everything that he had made, and behold, *it was very good*. And there was evening and there was morning, the sixth day (Genesis 1:31)." The short answer is that this statement was before the Fall when there was no death from PPPs. Just before God's five-star self-evaluation of the creation in Genesis 1:31, He describes the food source of all His creatures:

> And God said, 'Behold, I have given you every plant yielding seed that is on the face of all the earth, and every tree with seed in its fruit. You shall have them for food. And to every beast of the earth and to every bird of the heavens and to everything that creeps on the earth, everything that has the breath of life, I have given every green plant for food.' And it was so (Genesis 1:29–30).

From this passage we see that in the world before the Fall, predator-prey, parasite-host, and pathogen-host relationships were not part of the original ecology. After the Fall and the Flood, God opens up the menu to man by letting him kill and eat animals: "Every moving thing that lives shall be food for you. And as I gave you the green plants, I give you everything (Genesis 9:3)." Many questions may arise at this point but I hope to answer them in turn.

The curse was a judgment on Adam's sin which included mortality along with all sorts of unpleasantness—sweaty, backbreaking toil, thorns and thistles, pain in childbirth, and eventually the death of man (Genesis 3:16–19). Death would occur through murder, capital punishment, war (the sword), disease (pestilence), famine, predation (wild beasts), or old age. Both the Old Testament and New Testament clearly teach that death is an enemy and a consequence of our sin. It was not a part of the initial created order. "Behold, all souls are mine; the soul of the father as well as the soul of the son is mine: *the soul who sins shall die*

(Ezekiel 18:4, emphasis added)." "The wages of sin is death (Romans 6:23)." "The last enemy to be destroyed is death (I Corinthians 15:26)." It is clear from Scripture that this isn't just spiritual death nor is it confined to mankind.

> For the *creation was subjected to futility*, not willingly, but because of him who subjected it, in hope that the *creation itself will be set free from its bondage to corruption* and obtain the freedom of the glory of the children of God. For we know that the *whole creation has been groaning together in the pains of childbirth until now*. And not only the creation, but we ourselves, who have the firstfruits of the Spirit, groan inwardly as we wait eagerly for adoption as sons, the redemption of our bodies (Romans 8:20–23, emphasis added).

There are many other passages, but these will suffice. If you would like a more detailed case, I recommend an article by a world-class Hebraist, Dr. Pete Williams (Principal of Tyndale House, University of Cambridge) called "Genesis: The No-Agony-Before-Adam View."[43]

So the big question is this: Since God created each "kind" in an un-fallen state, how do we explain the vast array of biological badness: predators, parasites, and pathogens that were created prior to sin and death entering the world having complex anatomical, biochemical, and behavioral weaponry which deal out disease and death?

First, I want to establish scriptural boundaries outside of which we must not stray.

---

[43] Pete Williams, "The No-Agony-Before-Adam View," September 22, 2005, https://homeschoolscience.org/wp-content/uploads/2021/07/No-Agony-Before-Adam-Pete-Williams.pdf (accessed May 3, 2023)

1. God saw all that He had made, and it was *very good*.
2. There was *no (physical) human or animal death or disease* before the Fall. In other words, there were no predator-prey, parasite or pathogen-host relationships (Gen. 1:29–30, 2:16–17; Rom. 6:23, 8:19–22).
3. After the sixth day, God *finished* the work of creation (Gen. 2:1–2). I've already discussed the first two; however, this last one I believe eliminates the possibility that God did a creative act after the Fall by converting many creatures into harmful or death-dealing ones thus making the world a dangerous place.

In some cases the deadliness of certain pathogens and parasites can be attributed to degenerative mutations which consequently renders them lethal. In some cases the deadliness of certain predators could simply be a change in behavior. However, in many PPPs, this explanation is not satisfactory because they have such sophisticated anatomical and physiological weaponry that simply defies the idea that non-teleological processes like mutation or natural selection could cobble together such complexity (see Chapter 5). In the next section, I would like to unveil some of this weaponry by taking a closer look at one predator, one parasite, and one pathogen so the reader can see that these are way beyond the reach of unguided evolution. After that, I will discuss several scenarios that attempt to explain this biological complexity while embracing God's goodness and remaining within the confines of Scripture mentioned in the three points above.

DESIGNED TO DEAL OUT DEATH

Pit Vipers (Predator)

Few creatures inspire fear like the pit vipers with their strike speed and their long, curved, hinged front fangs that can instantaneously inject a powerful venom that consists of a number of nasty enzymes. They have killed thousands of people worldwide, injured many more, and killed countless millions of prey animals. Is this simply due to degenerative microevolution after the Fall? No. Let's look at the sophistication of the skull bones and muscles that swing the fangs out for venom injection. The fangs are mounted on the *maxilla bones* and lie on the roof of the mouth when at rest. During the strike, the *levator pterygoidei* muscles contract which shift the upper jaw bones forward. These bones pivot the maxilla bones like a see-saw thus swinging the fangs down and forward. The *ectopterygoid bone* serves as a fulcrum and a ligament braces the other end of the maxilla. As the head of the viper is hurled forward, the forward-pointing fangs deeply penetrate the prey's flesh. The fangs are curved hypodermic needles (Figure 36B). So where does the venom come from? The *venom glands* are large sacks situated on the outside and rear of the upper jaw bones (Figure 36A).

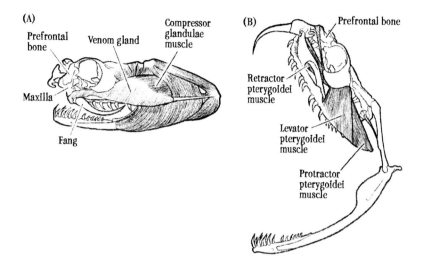

Figure 36: Viper venom injection apparatus.

The inside lining of the venom glands produces a cocktail of enzymes that wreak havoc on tissue. The multiple venom components vary in identity, activity, and quantity thus making a wide range of toxicities among the many viper species. Here are a few of these components:

- Peptide bradykinin potentiators—dilates and increases permeability of blood vessels
- Proteolytic enzymes—digests proteins or polypeptides
- L-amino acid oxidase—causes great tissue destruction
- Hyaluronidase—reduces the viscosity in connective tissue, increasing its permeability

- Phospholipase—digests cell membranes
- Phosphatases—destroys high-energy phosphate compounds (ATP)

We know this cocktail is injected into the victim, but how? The *compressor glandulae* muscle contracts, squeezing the venom gland like your thumb squeezing a tube of toothpaste. The venom rushes through the *venom duct* which leads to the hollow interior of the fangs. The venom courses through the fangs, out the opening near the tip, and deep into the flesh of the prey. The *retractor pterygoidei* muscles contract to shift the upper jaw bones backward, thus pivoting the maxilla so that the fangs swing back to the roof of the mouth.

There are other subtle design features to this apparatus, but you get the idea. Everything about this is designed. Just like the sophisticated design of the AK-47 rifle, there is no way to get around the fact that it is *designed* to be *deadly*.

Parasitoid Wasps (Parasites)

There are thousands of different species of parasitoid wasps spanning several different families of Order Hymenoptera (bees, wasps, and ants). They all have the insidious habit of laying eggs in other arthropods. I picked these wasps in particular because Darwin could not reconcile the idea of a benevolent creator purposely designing such horrific creatures. He found it more satisfying to attribute these creatures to naturalistic, evolutionary processes rather than to God's deliberate design. His theodicy dilemma was brought to a head when considering certain members of the Family

Ichneumonidae. Their females lay eggs in living caterpillars. The larvae consume their living larder from the inside out. Interestingly, this particular instance of *caterpillar suffering* appeared to unsettle Darwin even more than the parasites that cause much *human suffering*. Nevertheless he resolved this dilemma by simply making God not responsible for any of it, be it insect or human suffering.

I will describe a wasp of the genus *Glyptapanteles* of the Family Braconidae (similar to Family Ichneumonidae) and you will see why Darwin thought it problematic for a good God to make these dastardly creatures. This wasp injects its eggs into certain caterpillars. The larval wasps hatch and begin to feed on the hemolymph (nutrient rich blood) while instinctively avoiding vital organs of the caterpillar. The caterpillar continues to feed and grow while up to 80 wasp larvae grow within its body. When the baby wasps have finished their larval development, they exit by rasping an exit hole in the skin of the caterpillar with their tiny sharp teeth. Each larva spins their own little cocoon before they pupate. The larvae also succeed in mentally deranging the caterpillar so that it is semi-paralyzed and ceases to feed. Instead of spinning its own cocoon, it spins a blanket over the wasp cocoons adding another layer of silk to protect them against other parasitic wasps that are targeting them. The perforated, deranged, and dying caterpillar then defends the wasp cocoons by thrashing wildly to ward off any attackers until it dies.

Yersinia pestis (Pathogen)

There are many species of pathogenic bacteria and viruses, but none can hold a candle to the devastation wrought by one species of bacteria throughout world history, *Yersinia pestis*. It has killed roughly 300 million people spanning three major pandemics and a number of epidemics throughout world history. Unlike some predators and parasites, its appearance isn't scary; it looks like many other rod-shaped pathogenic and non-pathogenic bacteria. Its lethality is due to a suite of biochemicals called *Yersinia* outer proteins *(Yops)* that are injected into its host's cells using the *Type III Secretion System*. This is a complex injection apparatus composed of about 30 different proteins especially designed to pump a number of Yops into our cells. Two Yops form a molecular-sized pore through which the Type III Secretion System injects other Yops. These other Yops wreak havoc inside our cells, particularly important cells in our immune system. Some Yops disrupt the ability of our white blood cells to attack, engulf, and digest these deadly germs, and they also inhibit other functions important in the immune response. To use a Biblical phrase, they "bind the strong man." Another formidable feature of *Yersinia pestis* is "a cloaking device" of sorts. Flagella are molecular outboard motors mounted on the surface of the bacterium which enable it to swim. Flagella are very potent antigens (cell surface molecules) that alert our immune system to a very clear and present danger. The "cloaking device" is the bacterium's ability to stop making flagella once inside a warm-blooded animal. This allows it to evade and quickly outflank our immune system much better than other pathogens, often causing death before our body can adequately mount an effective immune response. It is also capable of enlisting certain species of fleas to effectively transmit itself from one infected mammal to another (rodent to rodent, rodent to human, or human to human). The bacteria-flea interactions are quite complex and clever but I will forgo the details for the sake of brevity. There are other features of *Yersinia pestis* that enable it to create all the life-threatening symptoms of the three forms of this disease—bubonic, septicemic, and pneumonic plague—but I have covered the main factors that make this pathogen so potent.

In the following section, I will lay out several scenarios that try to explain PPPs in light of Scripture and the biological facts. Recall our scriptural boundaries as we discuss these scenarios:

1. The creation was very good (Gen. 1:31).

2. Animals and humans were herbivores before the Fall and curse (Gen. 1:29–30) and there was no pestilence. Therefore, there was no animal or human death before the Fall (Rom. 6:23 & 8:19–22).[44]

3. God finished His work of creation (Gen. 2:1–2).

---

[44] Pete Williams, "The No-Agony-Before-Adam View," September 22, 2005, https://homeschoolscience.org/wp-content/uploads/2021/07/No-Agony-Before-Adam-Pete-Williams.pdf (accessed May 3, 2023)

SCENARIOS

In many cases, various creatures and plants have become pests or are simply undesirable because they have made themselves a nuisance in some way. We shouldn't jump to conclusions and say that every creature we find undesirable must mean that they were specifically designed to be a pain in the neck after the Fall. We must carefully examine their anatomy and physiology to see if their features are exclusively designed to inflict pain, suffering, and death. If not, then we might just chalk up their pest status to poor dominion or their high reproductive rate or both. For example, an herbivore might become a huge pest simply because they consume massive amounts of our crops. But we shouldn't be surprised at huge numbers of herbivores when we subsidize their reproduction by offering free room and board. Because of the curse, predators prey on herbivores and consequently their reproductive rate has been designed to increase to compensate for their losses due to predation. We live in a fallen world where herbivores can be our enemy and predators can be our friends. We too quickly assume that beautifully designed "plowshares" are a result of the curse when they destroy our crops (say weevils, grasshoppers, caterpillars, etc.). We often see "swords" (say ichneumon wasps) as a great blessing when they destroy crop pests, when actually their biology is more twisted as a result of the Fall. Nevertheless, it is perfectly good and lawful to employ them for our benefit in a fallen world. However, some designs can have dual purposes. A hammer can pound in a nail (good use) or bash someone's head in (bad use). For example, bears can use their chops to graze on grass or to kill animals for food. Their teeth and digestive tract design is versatile and lends itself to multiple uses. In this case, their fallenness is found in their *behavior* not in their *anatomy and physiology*. However, as mentioned in the previous section, there are many animals whose anatomy and physiology seem to be exclusively designed to deal out pain, suffering, and death. Besides using venom to produce antivenom for snake bite victims, it is hard to imagine a benign use for viper fangs, venom glands, and venom. The following scenarios are meant to address those features that only seem to fit into the economy of a fallen world.

Satanic Modification

This scenario envisions that the soon-to-be PPPs were good in all respects but after the Fall were transformed into actual predators, parasites, and pathogens by Satan or other powers of darkness through genetic modification. The problem I see with this scenario is that some of the most amazing designs out there would be found in the weaponry of predators. I am loath to credit Satan for them. I also don't believe he is intelligent enough to design this at the genetic level. The biblical reason I don't think Satan modified these creatures is that in the last few chapters of Job God describes some of the attributes of His creatures, some of which are predators, i.e., the hawk, the eagle, and leviathan. Interestingly, His description of a few animals is God's theodicy—His own justification of why He can allow bad things to happen to good people. In showcasing these creatures, He never gives Satan credit for their predatory features. Speaking of the eagle, "His young ones suck up blood, and where the slain are, there is he" (Job 39:30). Speaking of leviathan, "Who can open the doors of his face? Around his teeth is terror" (Job 41:14). In Job 42:2, Job exclaims, "I know that you can do all things, and that no purpose of yours can be thwarted." This implies that God is the Author of it all, including the formidable attributes of these creatures.

Macroevolution

This scenario envisions that the soon-to-be PPPs were good in all respects but after the Fall, naturalistic (unguided) processes of mutation and natural selection transformed them into actual PPPs. The huge problem with this is that mindless processes like random mutation and natural selection simply can't design this sophisticated weaponry. Non-intelligent processes cannot produce genetic information. Intelligent design and skill are required to beat plowshares into swords (see Chapter 4).

Divine Modification (post-Fall)

This scenario envisions that the soon-to-be PPPs were good in all respects but after the Fall, God transformed them into actual predators, parasites, and pathogens. The problem with this scenario is that, although God is still credited with creating this weaponry, it shifts it to a post-Fall creative act. This is problematic because of the fact that

God finished His work of creation on day six (Gen. 2:1–2). It appears the kitchen was closed, and therefore there was no more cooking up more biological designs. This scenario also suggests that God did not anticipate the Fall and had to scramble to design a Fall contingency plan after Adam sinned. But we know from Scripture that the Fall was preordained to happen; it was no surprise. The Lamb was slain before the foundation of the earth!

Divine Provision (pre-Fall)

This scenario envisions that the soon-to-be predators, parasites, and pathogens were created by God (genetically provisioned) for two ways of life—one for good structure, function, and behavior (sinless contingency plan) and one for predatory, parasitic, or pathogenic structure, function, and behavior (Fall contingency plan) with only the sinless contingency genes expressed prior to the Fall. After sin entered the world, the Fall contingency genes were expressed transforming herbivorous creatures into predators, parasites, and pathogens. To our modern ears, this may seem hard to fathom. How could snakes be completely herbivorous before the Fall? This shouldn't be a stretch for Christians who believe God has raised the dead. It is actually easy to envision a herbivorous creature transforming into a predator because we have actually seen it. In fact we are able to see it every spring if we want to. Consider the changes that occur when a tadpole becomes a frog. Their body undergoes a wholesale transformation, radically changing (both inside and out) from a legless teardrop-shaped fish-like herbivore with gills to a four-legged predator without gills. Their mouth and jaw change, their feeding behavior changes, their gut changes, their digestive enzymes change, and they grow legs, to mention a few. We all are familiar with this metamorphosis and think nothing of it because it happens all the time. Yet we boggle at the thought of herbivorous snakes transforming into predatory snakes. Why do we think of this as a fanciful fairy tale simply because we don't see it today? Oh, ye of little faith! View this scenario in light of the scripture, "For the creation was subjected to futility, not willingly, but because of him who subjected it, in hope that the creation itself will be set free from its bondage to corruption and obtain the freedom of the glory of the children of God" (Rom. 8:20–21).

I believe this fourth scenario is the best. It is in keeping with all scriptural boundaries laid out at the beginning of this chapter. It assumes the creation was very good before the Fall (Gen. 1:31). It assumes no human or animal death before the Fall (Rom. 8:20). And yet it assumes the *finished work* of creation on day 6 (Gen. 2:1). God doesn't have to step in and modify many creatures with some kind of post-Fall creative act. He has already endowed them with the genetic capacity that enables them to transform into either predators, parasites, or pathogens at the time of the Fall. These transformations are not evolutionary changes in any Darwinian sense; they are developmental changes due to differential gene expression of already created genetic information. In other words, certain genes are turned off and certain genes are turned on which produce these transformations. This is also one way of looking at the passage on thorns and thistles (which are clearly stated consequences of the curse in Gen. 3:18): "thorns and thistles it [the ground] shall bring forth for you; and you shall eat the plants of the field."

Rather than God creating thorns and thistles after He already created plants, He provisioned certain plants with the genetic capacity to produce thorns and thistles. Thus the ground brought them forth as part of the curse. Some might argue that if God provisioned certain creatures and plants to transform into death-dealing predators and thorns and thistles (at the Fall), why would He still call them *very good*? The simple answer is that the wrath of God against man's sin is *very good* because it is a manifestation of His moral perfection in holiness, justice, and judgment. The wages of sin is death (Rom. 6:23). Since man (who is the *imago Dei*) was given dominion over all creatures (Gen. 1:28), it makes sense that the consequence of the Fall extends to all creatures under his dominion.

There are other possible scenarios but they are slight variations of the above four. This thorny topic (no pun intended) is not often addressed in creation-evolution discussions, which is why I am discussing it here. The ideas presented in this chapter answered my initial question in a way that is harmonious with both Scripture and biological reality. Whether it is correct or not, I don't know. If not, I hope it will at least stimulate fruitful discussion that will move us toward the truth.

OPENING A CAN OF WORMS

If you assume I am correct about biological natural evil and if you follow this view to its logical conclusion you may have already discovered two seemingly untenable situations. First, if there is no animal death, you might immediately realize that all the world's creatures would hit an exponential growth curve that would make the phrase "population explosion" an understatement. Even with species with a low reproductive rate like elephants, we would be wall-to-wall elephants in no time. The elephant population would overshoot carrying capacity very quickly resulting in starvation and death due to lack of resources. This objection is easily dealt with by pointing out that God (Who is infinitely smarter than the smartest ecologist) would have anticipated that contingency and made provision for it. If Adam and Eve had not sinned, the creation would not be cursed and therefore overcrowding and starvation would not occur. God would presumably build in biological feedback mechanisms that would slow and halt reproduction once each kind or species reached an ideal global population size. In an uncursed world, all degenerative aging processes and death would not occur. Immortality (perfect self-maintenance negating or reversing all genetic entropy) would prevail in an unfallen world.

The other seemingly untenable situation would be the incidental death of small creatures by big, hulking creatures and inanimate objects. Even if you assume a perfectly benign ecology in which all creatures are herbivores, there is still a huge disparity in size between a large sauropod like *Brachiosaurus* and little aphids. Even one mouthful of leaves by the former could be the demise of thousands of the latter. There are countless other minor events that would spell doom to tiny creatures—footsteps, rockfalls, tree falls, earthquakes, fire, and any other natural disaster. I also assume that in an unfallen world, there would be no life-threatening natural disasters or other physical threats. Some creationists (including Creation Ministries International) have considered this conundrum and resolved it by assuming insects and other invertebrates are creatures that are not *nephesh chayyah*[45] (because they don't breathe through nostrils and don't have red blood). In so doing, they attempt to make inadvertent trampling or eating of insects a non-issue regarding the doctrine of "no death before sin." But even if you agree with CMI and assume vertebrates are the only *nephesh chayyah*, there are plenty of diminutive vertebrates that are just as small or even smaller than many invertebrates. They are just as vulnerable to being rubbed out by non-predatory acts of nature. My answer to this is speculative, as any thinking would be in considering the mysterious nature of an unfallen creation. My current thinking is that God gave all creatures in this non-destructive world an ultra-sensitive perception (over long distances) granting all creatures an intuitive awareness of each other's position and movement. Uncanny agility, instantaneous response time, and innate cooperation would allow small creatures to make way for large creatures with plenty of lead time. We get a glimpse of this when watching starling murmurations or hummingbird aeronautics around their feeders. If we know it will happen in the new heavens

---

[45] David Pitman. 2014. Nephesh chayyah: A matter of life...and non-life, https://creation.com/nephesh-chayyah

and the new earth, why couldn't it happen between creatures in the pre-Fall creation?

In Isaiah 11:6–9, the prophet looks forward to a redeemed creation in which there is no harm between man and animals and between different animals that currently are adversaries:

> The wolf shall dwell with the lamb,
>   and the leopard shall lie down with the young goat,
> and the calf and the lion and the fattened calf together;
>   and a little child shall lead them.
> The cow and the bear shall graze;
>   their young shall lie down together;
>   and the lion shall eat straw like the ox.
> The nursing child shall play over the hole of the cobra,
>   and the weaned child shall put his hand on the adder's den.
> They shall not hurt or destroy
>   in all my holy mountain;
> for the earth shall be full of the knowledge of the Lord
>   as the waters cover the sea.

Similarly, the Apostle Paul writes in Romans 8:20–21: "For the creation was subjected to futility, not willingly, but because of him who subjected it, in hope that the creation itself will be set free from its bondage to corruption and obtain the freedom of the glory of the children of God." The Apostle John writes with the same hope in Revelation 21:4, "He will wipe away every tear from their eyes, and death shall be no more, neither shall there be mourning, nor crying, nor pain anymore, for the former things have passed away."

CHAPTER 12
Tying Off Loose Ends

My goal in this book isn't to answer every possible question, but rather to proclaim the trustworthiness of Genesis and to demolish arguments and every lofty opinion raised against the knowledge of God. Darwinism is indeed raised against the knowledge of God, and is therefore my primary target. In addressing a limited number of topics, I wanted to get to the crux of each issue while avoiding massive data dumping. And even though I can't address every question, I wanted to circle back around to briefly address a few common questions that often arise in the minds of both bible-believing Christians and skeptics regarding the historicity of Genesis as it relates to certain scientific findings. I also want to point you to other sources if you want to dig deeper. Hopefully, we can connect some dots along the way.

Here are some of the questions that I will briefly answer:

- What about the dinosaurs? Were they on the ark?
- What about dinosaur soft tissue?

- How does the biblical timeline compare with the geological time scale?
- Do creationists believe in plate tectonics?
- What about the ice age?
- How did animals and plants get to different continents from the mountains of Ararat?
- What about star light arriving here when they may be millions or billions of light years away? Doesn't that prove an old universe?

There are many, many more possible questions, especially if you're informed and curious about the controversy. If I don't address your questions, I refer you to several good websites that offer a wealth of information that can help answer those questions on these and numerous other topics:

- Answers in Genesis (answersingenesis.org)
- Creation Ministries International (creation.com)
- Institute for Creation Research (icr.org)

Even though I commend my book and the above resources to you, that is not to say that everything you read in this book or from these organizations is Gospel truth. Even though we are trying to be faithful to Scripture and to scientific facts, we are fallible. Therefore, you, the reader, must be a Berean and search the Scriptures daily to see if these are true. Similarly, we should critically (and graciously) evaluate other creationists' interpretation of scientific data.

What about the dinosaurs?

The dinosaurs (as far as we know) are extinct. They were created on day six (at least the land ones). Completely aquatic dinosaurs and pterosaurs were created on day five. They coexisted with man.

Were they on the ark?

Yes. Just like all land animals and flying creatures, those terrestrial dinosaurs and pterosaurs were brought on the ark (male and female). They came off the ark alive but for unknown reasons went extinct sometime after the Flood. We can speculate as to why they died out, but it is imprudent to be dogmatic about our answers. For the record, I tend to think that over-hunting (for meat, rite-of-passage, wanton slaughter, etc.) coupled with rapid climate changes triggered by the flood and its aftermath contributed to their extirpation and ultimate extinction; but who actually knows, since it isn't mentioned in recorded history.

What about dinosaur soft tissue?

When Dr. Mary Schweitzer discovered real unfossilized soft tissue (collagen fibers, blood cells, bone cells, etc.) in a *Tyrannosaurus rex* femur in 2005, it shook up the scientific community worldwide. Many doubted it was actual dinosaur tissue simply because it was presumably 68 million years old. She stuck to her guns and shot down the doubters with impressive research showing that it was indeed dinosaur tissue and not, as some suggested, some other organic material that permeated the femur. The problem with this is that biological molecules (such as proteins, etc.) don't last anywhere near that long, even under ideal conditions. Even when deeply embedded in rock and cloistered away from decomposers like bacteria and fungi, biological molecules undergo decay due simply to heat energy within the bone or whatever. Even under very cold conditions, scientists estimated that they could possibly remain intact from one to three million years. However, most dinosaur bonebeds aren't kept in a deep freeze; they undergo many temperature fluctuations each year which are not conditions friendly to biomolecules. Those that believed it to be real tissue that is really that old needed to find a satisfactory explanation. The current theory is that iron from hemoglobin from a blood puree acts like a preservative on proteins. Research has been done demonstrating that bird blood vessels do not rot for 2 years at room temperature if soaked in this iron rich brew. So iron does have a preserving effect of the order of at least a couple years. Great! However, there are several findings that refute the iron preservation hypothesis, which you can read about in "Can Iron Preserve Fossil Proteins for Eons?"[46]

Though it's interesting biochemistry, it does not demonstrate how dinosaur collagen (and other proteins) can remain intact for tens of millions of years. The most parsimonious explanation is simply that dinosaur soft tis-

---

46 Brian Thomas, "Can Iron Preserve Fossil Proteins for Eons?" Institute for Creation Research, June 23, 2015, https://www.icr.org/article/can-iron-preserve-fossil-proteins-for/.

sue and its associated proteins are actually only a few thousand years old.

How does the biblical timeline compare with the geological time scale?

The secular view of Earth history (with its geologic time scale) is radically different from the biblical view of Earth history (with its time frame). This difference can be an occasion for embarrassment among the young-Earth creationists, especially for those who aren't science geeks and yet want to retain as much academic respectability as possible. Hopefully, this book will give you confidence while interacting with an unbeliever who may look at you squinty-eyed as if you're a member of the Flat Earth Society or the Alien Abduction Survivors Network. So, what is the difference? The secular view holds that the universe is about 13.8 billion years old, the earth is about 4.5 billion years old, and life on Earth is about 3.7 billion years old. Although there are a few mysterious multicellular life forms (e.g., Ediacaran fauna) late in the Precambrian era (just prior to the Cambrian Period), naturalists believe complex multicellular life supposedly evolved at the beginning of the Cambrian Period about 541 million years ago. The *Paleozoic Era* includes the Cambrian, Ordovician, Silurian, Devonian, Carboniferous, and Permian Periods and ends at about 251 million years ago. The *Mesozoic Era* (Triassic, Jurassic, and Cretaceous Periods) is referred to as the age of the dinosaurs and spans from about 251 to 65 million years ago. The most recent *Cenozoic Era* spans from 65 million years ago to the present. These immense time spans are referred to as *deep time*. Correlating the biblical timeline with the geologic time scale reveals radical differences. Most creationists believe the Precambrian rocks are a combination of "*ex nihilo* creation rocks" overlaid with non-fossil bearing sedimentary rocks. The latter may have been formed due to the massive erosion of sediment from the waters rushing off the new land masses that were rising from the waters when God said "let dry land appear" on day 3.

All the billions upon billions of fossils contained within the sedimentary rocks of the *Paleozoic* and *Mesozoic Eras*, presumed to span the time between 541 to 65 million years ago (amounting to about 476 million years of geologic time), were actually formed from about **one year** of sediment deposition during Noah's Flood! See what I mean by a radically different paradigm? Most creationists believe the *Cenozoic Era's* fossil-bearing rocks (Tertiary and Quaternary Periods spanning 65 million years ago to the present) are sedimentary layers deposited by smaller regional or local floods that occurred episodically after Noah's Flood. It is true that *Cenozoic* sedimentary rock units are much more limited geographically. The *Paleozoic* and *Mesozoic* rocks are much more extensive (spanning multiple continents) and can be more readily explained by the global scope of Noah's Flood. In short, the fossil record reveals the geographical, ecological, and elevational differences of various ecosystems each housing a unique array of organisms. These differences determine where and in what order the organisms were buried during this one global, catastrophic event called Noah's Flood that lasted about one year. Although we can speculate about what happened during the Flood, keep in mind that we are on skinny branches and it behooves us not to dogmatically jump up and down on them. Even if our speculations fit the data (at present) and the Bible, they could still be flat wrong. How can we honestly reconstruct the details of Noah's Flood when there is so much we simply don't know about the initial conditions and we weren't there to observe it happening? It's like trying to figure out the placement of each vegetable on the cutting board along with when and how fast they were chopped up and thrown into the pot…by only looking at the ingredients of vegetable soup after it's cooked and served.

Do creationists believe in plate tectonics and does it relate to Noah's Flood?

Most creationist researchers (but not all) currently accept the catastrophic plate tectonics (CPT) model as being causal to Noah's Flood. This model is quite powerful in explaining many geologic phenomena at least as well as the conventional plate tectonics model. And in some cases, it explains things better than the conventional model. The short version of CPT is that the supercontinent *Pangaea* broke up into two supercontinents *Gondwanaland* and *Laurasia*, and they in turn eventually broke up into the continents of today. Both moved around the globe in much the same way as the secular view suggests. The major difference is that in the CPT model, the oceanic and continental plates accelerate through the process called runaway subduction. Eventually, they reach speeds of meters per second—not millimeters per year! The commencement of CPT also explains some of the physical (secondary) causes God may have employed to bring on the flood waters described in Genesis 7:11–12.

For instance, when the initial break-up between oceanic plates occurred along all the mid-ocean ridges, it brought the ocean water in contact with the molten mantle. This would produce a series of steam jets thousands of miles long issuing from the fracture zones. This phenomenon may explain the phrase "all the fountains of the great deep burst forth" (Gen. 7:11). As all this steam shot high up into the atmosphere, it condensed into global-scale weather systems that would produce prolonged and intense global precipitation. This would explain the passage, "The windows of the heavens were opened. And rain fell upon the earth forty days and forty nights" (Gen. 7:11). As Genesis 7:19 says, "And the waters prevailed so mightily on the earth that all the high mountains under the whole heaven were covered." This verse clearly precludes the idea of a local flood. According to the current creationist view, most of the high mountains that exist today were a product of flood sediments that were heaved up during the continental collisions that occurred during and after the flood. The speed of CPT is also much better at explaining the force required to heave up a number of mountain ranges like the Himalayas (which, by the way, are loaded with marine fossils.) The mountains that existed prior to the flood were probably created when God made the dry land appear on day 3. There is much more nuance and detail to this model. As I said before, these brief answers are just scratching the surface to several questions but hopefully will stimulate your curiosity to dig deeper into the creationist literature. For a more thorough summary of this topic I refer you to *The New Creationism* by Paul Garner, specifically Chapter 13: "Global Catastrophe." The geophysicist who proposed and did most of the computer modeling of CPT is Dr. John Baumgardner. His model was so good even secular geophysicists used it, they just used a radically different time frame. He has written a more technical paper on this topic called "Catastrophic Plate Tectonics: The Physics Behind the Genesis Flood."[47]

What about the ice age?

Secular geologists believe that there were numerous ice ages that occurred over the course of 2.7 million years. A quick synopsis of the most widely held creationist view is that there was only one ice age that probably arose about 500 years (give or take a few centuries) after Noah's Flood and lasted some 700 years (again, give or take a few centuries). It is difficult to explain the onset of just one ice age let alone many. This is because the cold oceans in the polar regions don't produce the moisture needed to cause the heavy snow falls required to generate an ice age. However, the assumed geologic and atmospheric conditions that prevailed throughout the Flood were ideal for generating the conditions of an ice age. According to the most widely held creationist view of the Flood, oceans were relatively warm due to all the geologic activity of catastrophic plate tectonics (CPT). This, in turn, continued to cause massive amounts of evaporation and precipitation at a global scale even after the great Flood. In the temperate, arctic, and antarctic regions this precipitation would be predominately snow. To generate an ice age, the snowfall each winter

---

[47] John Baumgardner, "Catastrophic Plate Tectonics: The Physics Behind the Genesis Flood," *The Proceedings of the International Conference on Creationism*, 5 no. 1 (2003): 113, https://digitalcommons.cedarville.edu/icc_proceedings/vol5/iss1/13/

must exceed the snow melt each summer. Consequently, there would be a net accumulation of snow each year. Over many decades this will grow into continent-wide ice sheets and glaciers that advance from the high latitudes toward the lower latitudes. Many questions may arise in comparing this view with the secular view. A good summary of the evidence supporting one ice age can again be found in *The New Creationism* by Paul Garner in Chapter 15: "The Coming of the Ice." The most prolific creationist researcher on the ice age is Michael Oard. He has written multiple articles on the ice age that can be found online including "The Unique Post-Flood Ice Age."[48] If you want to dive deeper into this topic, I recommend his material.

How did animals and plants get to different continents from the mountains of Ararat?

This topic falls under the discipline of biogeography (the branch of biology that deals with the geographical distribution of plants and animals) which I mentioned previously in Chapter 10: "According to Their Kinds." Answering this question for plants from a biblical perspective is relatively easy. Through an enormous array of seed dispersal mechanisms, many plants are able to move across and between continents quite quickly. Also, many plants in seed form or through asexual reproduction were able to remain viable outside the ark, such as the olive (Genesis 8:11). Darwin actually proposed testable hypotheses regarding the ability

---

48 Michael Oard, "The Unique Post-Flood Ice Age," *Creation*, 41 no.4 (October 2019): 44-47, https://creation.com/ice-age-model

of seeds, plants, and certain animals to raft long distances in and on drifting mats of floating vegetation. His theory works equally well for creationists who need to explain the movement of some plants and animals across vast expanses of ocean waters. The rafting hypothesis may explain the biogeographical distribution of many plants and animals but some creationists seem to limit themselves to this mode of transport and ignore human agency. Why should we? After the Flood and Babel, it wasn't long before seafaring humans (probably the Phoenicians) were building boats and plying the oceans to explore this great big empty world. It seems reasonable to me that they would provision their boats with all sorts of plants and animals (for food and other purposes) for their long voyages. When successfully reaching a new continent or island, not only they but also the plants and animals they brought along, were able to colonize this new land accidentally or on purpose. Not everything can survive for months rafting intercontinentally on a soggy mat of vegetation. This is just another reasonable explanation that can be used in addition to the rafting hypothesis. Dispersal by human agents should not be underestimated.

Some might argue that many animals and plants are highly adapted to a narrow set of environmental conditions and they need eons of time to slowly evolve into their present narrow ecological niche. They might say, "No way could this highly specialized plant or animal survive its slow trek through extremely inhospitable habitats between the mountains of Ararat to some rainforest canopy of South America." The answer to this is that the highly specialized plant or animal species we see today is not what came off the ark. It is assumed that their ancestors (the ark kinds) were generalized physically and genetically. Therefore, they were highly resilient and adaptable so that they were able to cope with a variety of environmental conditions they might encounter as they disembarked and dispersed from the ark into a gnarly post-flood world. A combination of natural selection and epigenetic factors (acting over decades and centuries) presumably caused each ark kind to speciate into several (if not many) species, each becoming exquisitely adapted to the particular habitat they wound up in (see Chapter 10: "According to Their Kinds"). If you want to dig a little deeper into biogeography from a creationist perspective read the article "Biogeography" by Dominic Statham.[49]

What about star light arriving here when they may be millions of light years away? Doesn't it prove an old universe?

There have been a number of creationist theories proposed to explain the light arriving from distant stars in the biblical time frame. Some require invoking the theory of relativity and gravitational time dilation. Even though the people who formulated these models want to be faithful to the scriptures and to science, I feel they are trying too hard to explain a miracle in terms of theoretical physics. When discussing a miraculous event one doesn't need to explain it scientifically. If you can explain it scientifically, it isn't a

---

49 Dominic Statham, "Biogeography," *Journal of Creation*, 24 no. 1 (April 2010): 82-87, https://creation.com/biogeography

miracle. If God speaks the universe into existence *ex nihilo*, 6,000+ years ago, does He then say, "Oh dear, I forgot that light travels at a pokey 186,000 miles/second and those poor human folk I just made will have to wait for billions of years before the light from those stars can reach Earth 'to serve as signs to mark seasons and days and years, and let them be lights in the expanse of the sky to give light on the earth'" (Gen. 1:14–15). Surely the arm of the Lord is not too short. Is He flummoxed at getting light to its desired destination instantaneously? Why would tampering with the speed of light be insurmountable when God (disregarding the First and Second Laws of Thermodynamics) just instantaneously created everything out of nothing? When God miraculously makes distant galaxies, getting the light from there to here is part of the same miracle. Trying to make sense of a miracle scientifically is kind of funny. It's also a waste of time and effort. Why do we think God's miracle-making must cease immediately after the star is made and must be constrained by the laws of physics thereafter? The laws of physics, by the way, God also miraculously established. By making these assumptions, we are putting God in a box. Some say that God would be deceiving us by making anything with the appearance of age. Nonsense. By definition, anytime God does a miracle, He is making things behave in a way that doesn't make sense to cause-effect reasoning in a non-miraculous context. When Jesus made water into wine (miraculous context), would a wine taster be deceived into thinking that it was the product of the usually long process of good wine-making? Yes, he would, especially if he didn't know about Jesus' supernatural power. The normal sequence of events includes planting the grape seed, growing it into a mature vine, the time it takes for blossom growth, pollination, fruit growth, harvesting, grape pressing, fermenting, bottling, aging, etc. If you come to any miracle with uniformitarian assumptions, you risk charging God with deception which is something we should not be willing to do, especially during the miraculous creation of the heavens and the earth. "The distant starlight and time problem" is really a non-problem. You might as well say "the walking on water problem" or "rising from the dead problem." It doesn't prove an old universe any more than the very good wine served at the wedding at Cana proves that it must be at least 10–15 years old when it was minutes old, or that Adam and Eve were in their mid-twenties when actually they were a day or two old. Dr. Danny Faulkner, an astrophysicist on staff at Answers in Genesis, has a similar answer to the same question as I do. A popular-level article he co-wrote with Bodie Hodge is called, "What About Distant Starlight Models?"[50]

---

50 Danny Faulkner, "What about Distant Starlight Models?", Answers in Genesis website, accessed February 22, 2020 https://answersingenesis.org/astronomy/starlight/what-about-distant-starlight-models/

CHAPTER 13
The End of the Matter

As I come in for a landing, I want to summarize the integrity of the young-Earth (and universe) creation position and the utter failure of all the major pillars of the Darwinian view. This I have laid out, to my satisfaction, in previous chapters. I also want to mention that although sincere believers can hold various compromised views on Genesis (by incorporating deep time or evolution or both), they can slowly or quickly weaken their faith (and the faith of others) in what Scripture teaches on other important doctrines. Young-Earth creation isn't simply one acceptable view of Genesis in the smorgasbord of other equally acceptable views. This doctrine sets the foundation for all other doctrines. If we get this wrong but try to stay orthodox on other doctrines, the explanations get highly convoluted and untenable. If Christians hold to the secular narrative regarding deep time or evolution, then the doctrines relating to the existence of Adam (or where he falls on the timeline), the Fall, the curse, physical death as a consequence of sin, etc., are in danger of being offered up as a burnt offering on the altar of science or cast into weird shapes according to the current mold of science. In other words, they are either sacrificed or forced to fit the scientific narrative. Nonetheless, a commitment to biblical creation isn't something to beat other believers or unbelievers over the head with. We must stand firm while gently instructing those who are in error.

COURTESY IN CONTROVERSY

The following are biblical guidelines on how to deal with people (Christians or not, scientists or not) who differ with you a little or a lot on these matters. I have also included admonitions for creation scientists who get disgruntled at other creation scientists for unfavorable peer reviews and also admonitions to those who administer those unfavorable reviews.

It is apparent to me that division in the creation movement is widespread and at times there has been clear evidence of ungodly actions and sinful anger. Disagreement is not the main problem, discord is. In every scientific endeavor there will always arise strong disagreement between scientists, Christians included. Because our endeavors are to uphold biblical truth, some of our disagreements have also been of a theological nature. Disagreement can be healthy and good if all participants are walking in the light. It can lead to iron sharpening iron. However, it can also clearly lead to much sin if we allow those disagreements to sow discord among us. We want to do *good* science but we also want to do *godly* science. Although we may have opinions about certain creationist models and theories, this admonition is not in any way partisan. It is not about who is doing better creation science and who isn't. This is completely pastoral. I am well aware that the community of young-Earth creationists is a tiny fraction of the broader scientific community. It is also a member of the wider Christian community. Not only does it displease our Lord when we young-Earth creationists are not on speaking terms with each other, it is a poor testimony to the watching world. Will they know we are Christians by our love or by how staunchly we refuse to make things right? We think it is impossible to reconcile because it is impossible to get them to admit they're wrong and I'm right about a particular creationist model or ministry philosophy. I am sure the enemy is quite pleased at how well the root of bitterness is growing in the young-Earth creationist community over these years. To stop this growth, I want to address with Scripture two basic groups of folks—the offendees and the offenders. These groups include creation scientists as well as the creationist lay community.

To the Offendees

One of the reasons our relationships become strained (or broken) is that our research or paper was rejected by a reviewer. Remember, the merit or lack of merit of a paper is not my concern here. When my research is submitted for review, what should my attitude and response be to criticism or rejection? Here are some verses that come to mind. "Whoever loves discipline loves knowledge, but he who hates reproof is stupid" (Proverbs 12:1). Note that this verse is an indicative, not an imperative. A lover of knowledge loves discipline but stupid people hate correction. Even if we think the source of the discipline or correction (the reviewer) is wrong in his or her assessment, we should be humble, not hateful in receiving it. We don't want to fit the definition of a fool by how we respond. A little farther on in verse 16 it says, "The vexation of a fool is known at once, but the prudent ignores an insult." Even if the reviewer isn't manifesting Christian love but is outright insulting, wise people ignore or overlook it (water-off-a-duck's-back). Fools express their vexation immediately at an insult. In other words, they get huffy at correction. If we do, we are fools.

We must never hold a grudge because they didn't like our paper and were less than gentle in telling us as much. Our feelings and ego must not be woven into our scientific work. We must take our lumps patiently even if they were wrong in their criticism (in manner or content). We should either cover the offense or correct their manner. We must never gossip or nurse a grudge. If the reviewer's critique is obviously sinful (not just to you but to other neutral parties) and requires correction, then correction must be for their sake, not because our feelings were hurt and we want to retaliate. Galatians 6:1–2 says, "Brothers, if anyone is caught in any transgression, you who are spiritual should restore him in a spirit of gentleness. Keep watch on yourself, lest you too be tempted." What are the stipulations for correcting someone? First, you must be spiritual. That means not resentful or bitter at them. Love for them must be the motivation for the correction. Second, you must restore the person in a spirit of gentleness. You must not be vindictive or rancorous. "Let your gentleness be evident to all" (Philippians 4:5, NIV). We cannot believe we are upholding truth if we ignore God's truth about the way we are to go about it (2 Timothy 2:24–26). We need to remind ourselves that words like "gentleness" and "kindness" are every bit as biblical as the dogma we stand to defend and the models we earnestly believe in. This Christ-glorifying way of speech applies to those within and even those who oppose us outside our creation community. We must not react to each other or to the world like the world.

To the Offenders

If you fall into this camp, you might be pleased at my exhortation to the offendees. You might say to yourself, "Preach it, brother; it's about time someone told them to toughen up. The rigors of science are not for the faint of heart. It's an academic version of American football, after all." Here are some verses that come to mind. Again, Philippians 4:5 is a good rule of thumb not just for correcting sin but also criticizing science. A reviewer's gentleness must be evident to all. You must manifest a spirit of gentleness and must be spiritual (Galatians 6:1–2). You can still be honest and straightforward with your assessment of the science but it mustn't be barbed…or rash. Just because you are critiquing science, rather than correcting sin, doesn't mean you get to pull your gloves off. Proverbs 12:18 says, "There is one whose rash words are like sword thrusts, but the tongue of the wise brings healing." We can learn at least two things from this verse. First, rash (reckless) words do not necessarily have malicious intent, but they nonetheless have the capacity to do the same amount of damage. They pierce like a sword. Just because you didn't mean to hurt them doesn't mean you're not culpable. It does mean you're not wise. Second, we see that the opposite of a rash wounding tongue is a wise healing tongue. Ask yourself if your tongue is more apt to wound or heal people. These biblical principles don't go away when we interact with each other while doing creation science. We too often become numb to familiar verses like Ephesians 4:29: "Let no corrupting talk come out of your mouths, but only such as is good for building up, as fits the occasion, that it may give grace to those who hear." In saying all this, I am not advocating a low standard for science. We must maintain high standards in our science. But in holding that standard we must also love our brothers and sisters in the manner of critiquing each other's work. Our goal in all our speech (or writing) is to build each other up, to give grace to those who hear. If you feel that the paper you are reviewing is sub-par due to poor methodology or is contrary to your views, then communicate all of that clearly and graciously. There is no reason to be bitter at each other even if we sharply disagree. Paul and Barnabas experienced that and ministered separately for a time. I am sure that they didn't hold any animus toward each other. If they did, it would have been contrary to everything they were preaching.

Making Things Right

Whether you are an offendee or an offender or both, or a creationist lay person that is a loyal partisan of a particular offendee or offender, reconciliation is the order of the day. 1 John 1:7 says, "But if we walk in the light, as he is in the light, we have fellowship with one another, and the blood of Jesus his Son cleanses us from all sin." This is very clear. If we are walking in the light…we have fellowship with one another. It follows that if we do not have fellowship with one another, then one or both parties involved are not walking in the light. We may be scrupulous about the quality of science but are we as scrupulous about being in fellowship with one another? We often think fellowship is dependent on scientific agreement. No. True fellowship is based on all the parties involved walking in the light as He is in the light. Sometimes, we can have sweeter fellowship with someone in a different denomination than with someone in our own denomination. Why? Two people who have confessed their sins and have been washed clean are able to love, forgive, cover offenses, and consider others better than themselves. They can have fellowship with someone else in that same state of forgiveness, regardless of denominational or scientific differences. When we put on love, which binds everything together in perfect harmony, we are then able to discuss our differences in a spirit of unity and harmony. Ephesians 4:1–3 says, "I therefore, a prisoner for the Lord, urge you to walk in a manner worthy of the calling to which you have been called, with all humility and gentleness, with patience, bearing with one another in love, eager to maintain the unity of the Spirit in the bond of peace." We don't have to sweep our differences under the carpet. Colossians 3:16 says, "Let the word of Christ dwell in you richly, teaching and admonishing one another in all wisdom, singing psalms and hymns and spiritual songs, with thankfulness in your hearts to God."

In the gospels we can find the need for the offender to be urgent about seeking reconciliation with the offended (Matthew 5:23–26). We also find that the offended can approach the offender with a desire to win him or her (Matthew 18:15). The Holy Spirit, using the New Testament authors as a mouthpiece, gave us clear marching orders in these and many other passages. He has given us license to correct, admonish, and rebuke; but He also gave us clear instructions about how we should do it and the spirit in which we are to do it. We are to be gracious, kind, gentle, humble, forgiving, and eager to cover offenses.

If you are an offendee and are hurt, bitter, and resentful at another brother or sister (on a scientific or unscientific matter), you must forgive them. Even if they do not come to you for forgiveness, you are given the command to love, and Romans 12 gives us clear attitudes and actions that we, as Christians, must have. We should always love and always be ready to forgive. If it isn't there, you'll be slowly consumed with bitterness. If you know someone that has taken offense at your ungracious tone or rudeness or arrogant manner (even if you think your scientific assessment is correct), you must seek his or her forgiveness. If someone asks you for forgiveness, forgive completely and immediately (seventy times seven). Our prayer is to have the Holy Spirit hound all of us relentlessly until we lay our pride down, confess our sins, and reconcile (as far as it depends on us) ourselves to each other. God is doing a great work through the various creation research centers and the various creation ministries in spite of our sins against each other, lack of forgiveness, and discord. Creation ministry can be a great servant to the church, but to be so we must exhibit the same qualities that God expects from His church. They should know us by our love.

Can you imagine how much more God would use us if we were all in fellowship with Him and each other? Denominational and scientific disagreements seem like intractable differences, but His command that we be like-minded still stands (Philippians 2:2). Like-mindedness isn't the same as agreeably disagreeing…although it is a far cry better than disagreeably disagreeing. How can we strive toward like-mindedness if we refuse to love and forgive each other? Love and forgiveness is the only Christian way to bind everything together in perfect harmony. As in the case with Paul and Barnabas, our differences may preclude working in the same organization or research group (which may have different views, doctrinal statements, goals, etc.). However, those differences should never preclude true Christian fellowship (unless there's church discipline or apostasy involved) or give us license to nurse grudges, gossip, or fertilize the root of bitterness. For Jesus' sake, let us restore true Christian fellowship and forsake our petty grievances.

CONCLUSION
A Creationist Manifesto

Finally, I want to lay out a brief summary of this book. All Scripture is God-breathed and is useful for teaching, rebuking, correcting, and training in righteousness (2 Timothy 3:16). The Bible is true (from beginning to end) but must be interpreted carefully using hermeneutical principles appropriate for the genre. In other words, history should be interpreted as history, poetry as poetry, etc. Because Genesis is written as straightforward history, it shouldn't be reinterpreted into something contrary to what it obviously says even if mainstream secular scientific claims are contrary to its contents.

The earth is a little over six thousand years old based on the Hebrew meaning of the word *yom* and the genealogies of Genesis 5 and 11. Also, several New Testament writers speak of the events and people of Genesis as historical not figurative. Plants, animals (including dinosaurs), and people were created during the creation week which lasted six ordinary days followed by one day of rest to set a pattern for our work week (Genesis 1:1–Genesis 2:4).

Radiometric dating methods are unreliable and do not necessitate a reinterpretation of the clear historico-grammatical interpretation of Genesis.

All complex, specified information requires intelligence. This includes genetic information in all life forms. Darwinian mechanisms over any amount of time cannot produce it or add to it. In fact the genetic information created at the start cannot maintain itself indefinitely apart from God's intervention. It slowly degrades. Natural selection may slow its degradation, but it cannot stop it. It slowly and inexorably degenerates according to natural law.

The global Flood in the days of Noah destroyed all terrestrial animal life except for eight people and all terrestrial animal kinds brought into the ark (Genesis 6–8). This global event has left an enormous amount of geological and paleontological evidence which should be interpreted in light of Scripture. The clear gaps between human, animal, and plant groups in the fossil record highlight the biblical truth that God created the various kinds. It is difficult to determine, given our current knowledge, what those biblical kinds are with certainty.

At the end of the creation week, "God saw all that He had made and it was very good" (Genesis 1:31). It is clear from the context of Scripture that animal and human death was and is a result of the Fall—God's curse against Adam's disobedience (Genesis 2:17). All animals and humans were vegetarians before the Fall (Genesis 1:29–30). The Fall also resulted in a curse that affected the entire creation. We know that the whole creation has been groaning as in the pains of childbirth right up to the present time (Romans 8:22). This is manifested through thorns and thistles, hard and painful work for man, pain in childbirth, the transformation of benign creatures into predators, parasites, and pathogens, and the ultimate physical death of all creatures possessing *nephesh*.

Biological systems at all levels (molecular biology to ecology) are riddled with the consequences of the Fall, and it should be understood that even though they are glorious and wonderful in many respects, they aren't operating presently the way God originally created them. All creation will be utterly transformed into something too glorious for us to currently grasp. "The creation itself will be set free from its bondage to corruption and obtain the freedom of the glory of the children of God" (Romans 8:21). John echoes this eschatological hope. In Revelation 21:4 he writes, "He will wipe away every tear from their eyes, and death shall be no more, neither shall there be mourning, nor crying, nor pain anymore, for the former things have passed away."

GENERAL INDEX

abiogenesis, 100
Acanthostega, 69–70
Archaeopteryx, 56, 76–78, 83
Ardipithecus ramidus (ARDI), 87, 89–90
Australopithecus afarensis, 87, 91–93, 96–97
Australopithecus sediba, 88, 96
Augustine, St., 11

baramin(s), 150–157, 159–161, 165–166
baraminologist(s), 153, 156, 162
baraminology, 150, 153, 155, 162, 166
Barr, James, 2–3
Behe, Michael, 104, 118, 120
 Darwin's Black Box, 118
Big Bang, viii, 3
biogeographers, 162
biogeography, 153, 162, 196, 198

Cambrian Explosion, 62–63, 65
Cambrian Period, 61–62, 65, 68, 191

Carboniferous Explosion, 65–66
Carboniferous Period, 65–66, 191
catastrophism, 59
catastrophist(s), 59
Christian(s), ix, x, 1, 5, 8, 14, 30, 153, 167–168, 181, 187, 201–204, 208–209
Concerto Effect, 109, 111
confirmation bias, 88–89
Corner, E. J. H., 81
CPT (catastrophic plate tectonics), 193–195
created kind(s), 34–35, 66, 143–146, 150, 153–154, 159, 162–163, 165
creationism, 101
creationist(s), vii–viii, 14–15, 24, 29–31, 33, 49, 59, 71–72, 83, 89, 145, 147, 149–150, 152–153, 157, 165–166, 185, 188, 192–193, 197

    progressive, 1, 143

    old-Earth, 9

    orchard, 150–151

    young-Earth, 1, 7–8, 11, 25, 29–30, 59, 191, 203

Darwin, Charles, 55–56, 86, 119, 147–149, 174–175, 196

    The Origin of Species, 55–56, 100, 148

Darwinian, 53, 115, 129, 135, 137, 141, 182, 201, 212

    neo-Darwinian, 14

Darwinism, x–xi, 52–54, 130, 133, 187

    naturalistic, 119

Darwinist(s), vii, 52, 118, 137
deep time, 5, 7–9, 11, 13–15, 30, 191–192, 201
Devonian Period, 63, 67–69, 191

Earth, 14–15, 17, 30, 50, 59–61, 99, 105, 107–108, 111, 142, 169, 181, 186, 191, 194, 199–200, 211

age of the, 5, 7, 11, 13–14, 25
history, 7, 59, 191
planet, viii
young, 11, 25, 27, 162
Eldridge, Niles, 60, 64
emigration, 32, 34, 37
entropy, 136
genetic, 184
epigenetics, 31
Eusthenopteron, 69–71
evolution, vii, x, 13–15, 29–31, 36–37, 44–45, 48, 52, 54, 62–63, 71, 74, 77, 82–83, 85, 98, 100–101, 135–136, 171, 201
biological, 3
chemical, 105
Darwinian, 135, 137
gradualistic, 55–56, 60–61
human, 86–87, 90, 93, 98
model, 83
naturalistic, x
neo-Darwinian, 14
reductive, 95–97
theory of, 14, 81, 89, 101, 118
evolutionary, 96, 119
ancestor(s), 62
bias, 97
biologist(s), 89
change(s), 182
history, 119
mechanisms, 38

mutation(s), 36
paleontologist, 60
process/processes, 50, 174
progression, 40
scientist(s), viii
theory, 55–56, 96
thinking, 14, 43, 52, 101
time, 64
tree, 56, 74, 76, 88, 90, 100, 141, 145, 149–151, 153
evolutionist(s), vii–viii, 9, 29–31, 37, 41, 43–44, 77, 83, 88–89, 93, 101, 120, 145, 153, 156, 165
mechanistic, 104
secular, 145
theistic, 1, 143
ex nihilo, 192, 199

Fall, the, 168–169, 171–172, 177–183, 201, 212–213
fixity of species, 147–148
See also *Linnaean Lawn*
Flood, 59, 150, 169, 189, 192, 194–195, 197
global, 2, 11, 212
local, 192, 194
Noah's, 2, 59, 192–193, 195

gap theorist(s), 1
Garner, Paul, 194, 196
The New Creationism, 194, 196
gene flow, 32, 34, 49–50, 53, 141
genetic drift, 33, 38, 49–50, 53, 141
genetic recombination, 32, 35, 49–50, 53, 141, 153

genetic variation, 29–30, 32, 34, 38–39, 50, 54
See also *microevolution*
genome(s), 102, 135–137, 156–159, 165
genomic equivalence, 155–160, 165
geochronological, 27
geochronologist(s), 15, 17, 20–21, 23, 27
geochronology, 25
geologic/geological time scale, 25, 61, 162, 188, 191–192
geophysicist(s), 194–195
Gould, Stephen J., 60–61, 64

Haldane, J. B. S., 105
Homo erectus, 88, 93–95
Homo floresiensis, 88, 95
Homo habilis, 88, 92–93, 96
Homo naledi, 88, 97
Homo neanderthalensis, 88
Homo sapiens, 88, 91, 93–95
hybridization, 34, 153–154

ice age, 188, 195–196
Ichthyostega, 69–70, 72–73
imago Dei, 183
immigration, 32, 34
Information Theory, 44, 49, 133
inbreeding depression, 95–97
Intelligent Design (ID), vii
irreducible complexities/complexity, 104, 118, 120, 127–128, 133
isochron dating, 21–23

isochron method, 23

Linnaeus, Carolus, 147
Linnaean
   hierarchy, 146
   Lawn, 147–150
      See also *fixity of species*
   levels, 143
Lyell, Charles, 56
   Principles of Geology, 56, 59

macroevolution, 29, 34, 38–40, 48, 52–54, 135, 162, 180
macroevolutionary, 30
materialist(s), 109
materialistic, 42
Mayr, Ernst, 52–53
   What Evolution Is, 52
microevolution, 29–34, 38–39, 54, 135, 162
   See also *genetic variation*
Miller, Stanley, 107–108
morphology, 151, 153–155, 161
multicellular, 61–62, 191
mutation(s), x–xi, 33, 35–37, 41–42, 44–45, 48–50, 53, 62, 119, 130, 133, 135–138, 141, 150, 153, 157–158, 160, 171, 180

natural selection, x–xi, 33, 37–38, 45, 48–51, 53, 62, 135–138, 141, 148–149, 153, 157, 171, 190, 198, 212
neomorph(s), 39–40, 43, 48–50
neomorphic, 53
nephesh chayyah, 185, 213

Oard, Michael, 196
old-Earth
    interpretations, 8
    creationists, 9
    view, 11
Oparin, Alexander, 105
Oparin-Haldane Hypothesis, 105, 107
origin of life, 3, 100

paleoanthropology, 86
paleontological, 212
paleontologist(s), 60, 70, 77, 161
paleontology, 56, 153, 161
paleo-expert(s), 87–96, 98
plate tectonics, 188, 193
    catastrophic, 162, 194–195
        See also *CPT*
    gradual, 162
post-Fall, 180, 182
Precambrian, 61–63, 65, 191–192
pre-Fall, 181, 186
punctuated equilibrium, 64

radiometric date(s), 27
radiometric dating, 13–15, 21, 23–24, 27, 212
reductive evolution, 95–97

Sandford, John C., 136
    Genetic Entropy, 136
sarcopterygian(s), 69–71

Scripture, 1–3, 10, 83, 144–145, 170–171, 177, 181–183, 188, 198, 201, 203, 211–212
selection pressure, 38
shanah, 8
speciate(s), 74, 150, 198
speciation, 29, 149–150, 153, 160
Statham, Dominic, 198

taxonomic, 154–155
taxonomist(s), 161
taxonomy, 143, 147
theodicy, 167–168, 174, 179
Tiktaalik, 70–71, 77, 83

unicellular, 56, 61, 115, 120
uniformitarian(s), 27, 59, 200
uniformitarianism, 27, 58–59, 60
Urey, Harold 105, 107

Wheeler, William M., 155
Whewell, William, 58

yom, 6–7
young-Earth
    creation, 27, 201
    creationist(s), 1, 7–8, 11, 14, 25, 29–30, 59, 191, 203
    position, 1, 8, 201

Made in United States
Cleveland, OH
26 July 2025